零失敗
秘方系列

新手入廚
做好餸

Skilful recipes for novice cooks

編者話
Preface

入廚，放鬆心情！

　　走入廚房，絕對是一項紓緩平日緊張情緒的生活方法。尤其對初下廚的新鮮人來說，能夠完成一道滿意的菜餚或甜點，那份滿足感由心底裏散發出來。

　　給料理新手的，可嘗試：黑椒豬肉乾、甘蔗雞、雞茸豆腐球、鮮雜果杏仁豆腐……

　　喜歡挑戰高難度菜式的，可試做：黃金糯米釀雞翼、自製 XO 醬、燒腩仔、金盞海鮮燴、鳳梨酥卷……

　　書內介紹多款烹調小技巧，如切鴿片、自製魚腐、釀蝦膠、包魚茸餃、迷你月餅等，附詳細步驟圖片，任何菜式、任何技巧，絕對難不到你！

Relax! It's just the kitchen.

A cooking session sure is a way to release the stress of the busy daily life. This is especially true for those who are new to the kitchen. Successfully nailing a dish or a dessert gives them incomparable satisfaction that is hard to put down in words.

For the absolute beginners, try pork jerky with black pepper, chicken sugar cane skewers, deep fried chicken and tofu balls, almond tofu pudding with assorted fruits.

For those more adventurous, try chicken wings stuffed with salted egg yolk and glutinous rice, home-made XO sauce, roasted pork belly, braised seafood and zucchini in crispy cups, crumbly pineapple pastry rolls.

This book covers many basic cooking skills, such as slicing a squab, making tofu fish puffs from scratch, mincing shrimps, making minced fish dumplings and mini mooncakes. All steps are clearly illustrated so that you won't fret over any new recipe you've never tried or any new trick you've never done before. Enjoy!

目錄
Contents

燒腩仔 / 56
Roasted Pork Belly

芝心蝦丸 / 60
Cheese Stuffed Shrimp Balls

海南雞飯 / 63
Hainanese Chicken Rice

魚茸餃兩吃 / 66
Fish Dumplings in Two Flavours

叉燒醬豬腩肉雙拼 / 70
Grilled Pork Cheek and Belly in Char Siu Sauce

雞茸豆腐球 / 73
Deep Fried Chicken and Tofu Balls

甘蔗雞 / 76
Chicken Sugar Cane Skewers

露筍雞髀菇炒鴿片 / 79
Stir-fried Pigeon with Asparagus and King Oyster Mushroom

咕嚕百花釀油條 / 82
Shallow-fried Stuffed Dough Stick with Minced Prawn Paste

西式鹹甜點
Western Pie & Desserts

田園芝士豆腐批 / 84
Vegetables and Tofu Pie with Cheese

菠菜番茄火腿餡餅 / 86
Spinach and Tomato Quiche

芝士蝴蝶酥 / 89
Cheese Palmier

朱古力花生醬扭紋蛋糕 / 92
Peanut Butter Chocolate Marble Muffins

豆渣果仁脆曲奇 / 95
Soybean Pulp and Nut Cookies

中式甜點
Chinese Sweets

迷你藍莓果仁月餅 / 98
Mini Mooncakes with Blueberry and Assorted Nuts

鮮雜果杏仁豆腐 /102
Almond Tofu Pudding with Assorted Fruits

黑芝麻杏仁千層糕 / 106
Black Sesame and Almond Layer Cake

鳳梨酥卷 / 109
Crumbly Pineapple Pastry Rolls

打好入廚基本功，
裝備自己

Equip Yourself with
the Basic Cooking Techniques

入廚烹調並非難事，只要掌握基本的入廚小竅門，
你我都可以成為家中的大廚，輕鬆入廚。

Cooking is not difficult at all. As long as you grasp the kitchen basics, everyone can be a competent home cook without breaking a sweat.

鮮拆蟹肉無難度

Pick Fresh Crabmeat the Easy Way

蟹肉是常用的材料，自製鮮蟹肉鮮甜味美，而且方便快捷，快快動手吧！

◎◎ 做法

1. 蟹身及蟹鉗隔水蒸熟，待涼或冷藏 2 小時，容易拆出蟹肉。
2. 用刀略拍蟹鉗，拆出完整的蟹鉗肉。
3. 用剪刀剪出蟹身白殼，用匙羹容易刮出蟹肉。

◎◎ 入廚小竅門

☆ 蟹肉內的碎殼必須徹底撿出，以免刺傷口腔。

1

2

3

4

5

Crabmeat is a common ingredient. Self-made fresh crabmeat tastes sweet and is also easy and convenient to make. Let's do it now!

Method

1. Steam crab body and crab pincers until done. Set aside to let cool and refrigerate for 2 hours.

2. Crack the pincers gently with the flat side of a knife. Shell them and take the pincers out in one piece.

3. Cut the white shell on the crab body with a pair of scissors. Scoop out crabmeat with a spoon.

Cooking Tips

⭐ Any broken shell in the crabmeat must be picked and removed completely so that no one chokes on them.

自家製蝦膠
Making Minced Prawn from Scratch

◎ 做法

1. 蝦肉抹乾水分後放於砧板,用刀面大力拍扁蝦肉,再用刀面推開。
2. 用刀背在蝦肉略剁成蝦茸。
3. 放入深碗內,順一方向攪拌至起膠即成,或冷藏 1 小時後使用,效果更佳。

◎ 入廚小竅門

⭐ 打蝦膠時,碗內必須徹底乾淨,不可沾有蒜、薑等食材,否則難以打成蝦膠。

⭐ 可考慮使用冷藏蝦,容易打成蝦膠。

⭐ 順一方向攪拌後,可撻入碗內數次,可增加黏稠感。

◎ Method

1. Wipe dry shelled prawns and put on a chopping board. Crush them with the flat side of a knife forcefully. Smear them sideways with the flat side of a knife.

2. Chop shelled prawns with the back of a knife until fine.

3. Put minced prawn into a deep bowl. Stir in one direction until sticky. For the best result, refrigerate the minced prawn for 1 hour before use.

◎ Cooking Tips

⭐ The bowl for making minced prawn must be thoroughly clean and should not be contaminated with other ingredients like garlic and ginger. Otherwise, the minced prawn won't turn sticky.

⭐ You may use frozen prawns for this recipe. They tend to turn sticky and bouncy more quickly than fresh ones.

⭐ After stirring the minced prawns in one direction, lift the minced prawn and slap it back into the bowl forcefully a few times. That would make the minced prawn more sticky and bouncy in texture.

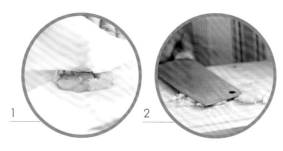

1

2

3

4

自製魚茸

Making Minced Fish from Scratch

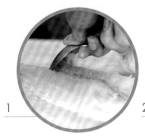   

1 2 3 4

做法

1. 用鐵匙順魚肉的逆紋刮出魚肉，用手檢查魚肉是否含有小骨。

2. 用刀剁成幼滑的魚茸。

3. 加入調味料攪拌，拌至帶黏稠性，下芫茜拌勻。

4. 將魚茸反覆多次搋入碗內，至魚茸不黏碗為宜。

5. 放入雪櫃冷藏片刻，令魚茸略收乾。

入廚小竅門

⭐ 必須小心檢查魚肉是否藏有小骨，以免吃魚茸時刺傷口腔！

⭐ 用多雙筷子頭粗端攪拌魚肉，容易攪拌均勻。

⭐ 用水弄濕雙手，以免魚茸黏着手。

Method

1. Scrape the meat off the skin and bones across the grain with a stainless steel spoon. Check whether the meat contains tiny bones.

2. Finely chop the meat until smooth.

3. Mix the minced fish with the marinade and stir until sticky. Add the coriander and give it a good stir.

4. Lift the minced fish mixture and slap it back into the bowl repeatedly until it does not stick to the bowl.

5. Refrigerate briefly to dry the minced fish.

Cooking Tips

⭐ All tiny bones must be removed so that no one chokes on them.

⭐ Stir the minced fish with the thick ends of a few pairs of chopsticks to speed up the process.

⭐ Keep your hands wet so that the minced fish won't stick to your hands.

雞鎚褪肉

Making Chicken Lollipop with a Drumstick

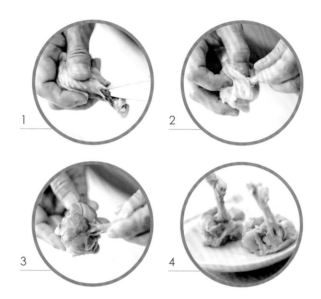

1
2
3
4

◎ 做法

1. 用剪刀沿雞鎚的骨位剪入，剪去雞肉黏附的筋部，肉與骨更易分離。

2. 用手將雞肉慢慢推到雞鎚頂端，弄成球狀即成。

◎ 入廚小竅門

⭐ 徹底將雞肉推到頂端，你會嘗到啖啖肉的雞鎚！

⭐ 此方法絕對不難，最重要是耐性及細心！

◎ Method

1. Insert scissors in between the skin and the flesh on the smaller ends of the drumstick. Cut the meat as closely to the bone as possible. Cut the tendon sticking to the meat and separate the meat from the bone.

2. Flip the meat inside out and upward towards the thicker end of the drumstick. Shape the meat like a ball.

◎ Cooking Tips

⭐ Push the meat completely upward and you get a lollipop with the bone as a convenient handle.

⭐ This method is absolutely easy as long as you're patient and careful.

雞全翼去骨

Deboning a Whole Chicken Wing

◯◯ 做法

1. 切去雞全翼的雞上翼部份。

2. 用剪刀沿雞中翼的骨位剪入，剪去雞肉黏附的筋部。

3. 沿雞骨向下慢慢褪出雞肉，最後拆出雞骨。

◯◯ 入廚小竅門

☆ 若雞肉黏着雞骨難以分拆開來，建議用小刀協助慢慢削出雞肉。

☆ 小心別剪破雞翼外皮，否則釀入的餡料容易溢出！

◯◯ Method

1. Cut off the drumette on the wing.

2. Insert scissors at the cut on the mid joint between the skin and the bone. Cut the meat as closely to the bones as possible. Cut any tendon that adheres to the meat.

3. Pull the meat toward the winglet end to expose the bones. Break and remove the bones.

◯◯ Cooking Tips

☆ If the meat sticks to the bones too firmly, use a small knife to scrape the meat off little by little.

☆ Keep the skin intact when you cut the meat from the bones. Otherwise, the filling stuffed inside the wing may leak when cooked.

1

2

3

4

瑤柱三文魚醬

Flaked Salted Salmon with Dried Scallops

◎ 材料

鹹三文魚 6 兩
乾瑤柱半兩
薑茸 2 湯匙
蒜茸 1 湯匙
米酒 2 湯匙
粟米油 1.5 杯

◎ 鹹三文魚做法

1. 急凍三文魚一塊解凍，洗淨，抹乾。

2. 取瓦瓷或塑膠容器，先鋪一層粗海鹽，放入魚扒，最後蓋滿粗海鹽。

3. 加蓋，密封，冷藏於雪櫃鮮果格 2 至 3 星期。

◎ 做法

1. 乾瑤柱洗淨，浸軟，撕成瑤柱絲。

2. 鹹三文魚隔水蒸 10 分鐘，待冷，拆肉弄散。

3. 燒熱鑊下油半杯，下瑤柱絲炒香，加入薑茸炒香，下三文魚肉及蒜茸用慢火炒透及壓碎，濺酒，邊炒邊加入餘下的油（油必須蓋過所有材料），再炒片刻，盛起，待冷入瓶儲存。

◎ 入廚小竅門

⭐ 瑤柱及三文魚必須用油炒至乾透，徹底去掉所有水分，令醬汁更美味，而且耐存。

⭐ 此醬放於雪櫃可儲存半年。

1

2

3

⓪ Ingredients

225 g salted salmon
19 g dried scallops
2 tbsp grated ginger
1 tbsp finely chopped garlic
2 tbsp rice wine
1.5 cups corn oil

⓪ Method for salted salmon

1. Thaw 1 piece of frozen salmon. Rinse well and wipe dry.

2. Put a layer of coarse sea salt on the bottom of a ceramic or plastic airtight container. Put in the salmon and top with more sea salt until full.

3. Cover the lid and seal well. Store in the crisper compartment of your fridge for 2 to 3 weeks.

⓪ Method

1. Rinse the dried scallops. Soak them in water until soft. Tear them apart into fine shreds.

2. Steam the salted salmon for 10 minutes. Leave it to cool. Skin and debone it. Break it into flakes.

3. Heat a wok and add 1/2 cup of oil. Put in the dried scallops and stir until fragrant. Add grated ginger and stir fry until fragrant. Add flaked salmon and garlic. Stir fry until done. Crush the salmon with a spatula. Sizzle with wine. Pour in the remaining oil while stirring continuously. There should be enough oil to cover all ingredients. Stir fry briefly. Set aside to let cool. Transfer into sterilized bottles.

⓪ Cooking Tips

⭐ Make sure you stir fry the dried scallops and salmon until thoroughly dry so as to remove all moisture. That would ensure its shelf life.

⭐ This sauce lasts well in the fridge for up to 6 months.

4

5

入廚技巧

XO 醬
XO Sauce

◎ 材料
乾瑤柱 1 兩
蝦乾 1 兩
金華火腿茸 1/4 兩
指天椒 1.5 兩
蒜茸 1 湯匙
乾葱茸半湯匙
粟米油 1.5 杯
米酒 1 湯匙

◎ 調味料
蠔油 1 湯匙
鹽 1 茶匙
糖半茶匙

◎ 做法
1. 乾瑤柱及蝦乾浸軟,瑤柱撕成絲;蝦乾切碎。
2. 指天椒洗淨,去蒂,切粒。
3. 燒熱半杯油,下乾瑤柱絲及蝦乾炒香,炒約 10 分鐘至起泡,潷酒,下乾葱茸、指天椒、蒜茸及金華火腿茸炒香,下調味料炒勻,注入餘下油分,煮至油滾,再煮片刻即成(撇去油面泡沫),待涼,入瓶儲存。

◎ 入廚小竅門
⭐ 瑤柱及蝦乾含水分,故必須炒至乾透,水分充份揮發掉,令醬料不容易發霉,保存質素。
⭐ 此醬放於雪櫃可儲存半年。

1

2

3

Ingredients

38 g dried scallops
38 g dried prawns
10 g grated Jinhua ham
57 g bird's eye chillies
1 tbsp finely chopped garlic
1/2 tbsp finely chopped shallot
1.5 cups corn oil
1 tbsp rice wine

Seasoning

1 tbsp oyster sauce
1 tsp salt
1/2 tsp sugar

Method

1. Soak the dried scallops and dried prawns in water until soft. Tear the dried scallops into shreds. Finely chop the dried prawns.

2. Rinse the bird's eye chillies. Cut off the stems and dice them.

3. Heat a wok and add 1/2 cup of oil. Stir fry dried scallops and dried prawns until fragrant. Keep on stirring for about 10 minutes until it bubbles. Sizzle with wine. Add shallot, bird's eye chillies, garlic and Jinhua ham. Stir fry until fragrant. Add seasoning and stir well. Pour in the remaining oil. Cook until the oil boils. Keep on cooking a bit longer. Skim off the foam on the surface. Leave it to cool. Store in sterilized bottles.

Cooking Tips

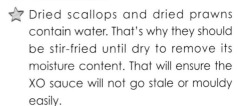

⭐ Dried scallops and dried prawns contain water. That's why they should be stir-fried until dry to remove its moisture content. That will ensure the XO sauce will not go stale or mouldy easily.

⭐ This sauce lasts well in the fridge for 6 months.

4

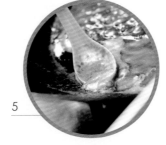

5

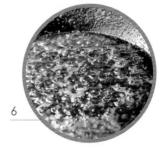

6

叉燒醬

Char Siu Sauce (Cantonese Barbecue Sauce)

◐ **材料**
葱白 6 段
蒜肉 6 粒
薑 4 片

◐ **調味料（拌勻）**
磨豉醬、紅糖醬、紹酒各 1 湯匙
生抽 2 湯匙
蠔油半湯匙
胡椒粉少許
麥芽糖及水各 3 湯匙

◐ **做法**

1. 燒熱鑊下油 2 湯匙，下葱白、蒜肉及薑片炒香。

2. 加入調味料，用慢火煮至濃稠，最後去掉葱白、蒜肉及薑片即成。

◐ **入廚小竅門**

⭐ 紅糖醬是紅麴用糯米及酒發酵而成，是一種天然健康的醬料。

⭐ 因含有麥芽糖，故烹調時需不斷攪拌，以免黏底。

⭐ 此醬放於雪櫃可儲存 3 日。

Ingredients

6 short lengths of white part of spring onion
6 cloves skinned garlic
4 slices ginger

Seasoning (mixed well)

1 tbsp ground soybean paste
1 tbsp red vinasse
1 tbsp Shaoxing wine
2 tbsp light soy sauce
1/2 tbsp oyster sauce
ground white pepper
3 tbsp maltose
3 tbsp water

Method

1. Heat a wok and add 2 tbsp of oil. Stir fry the spring onion, garlic and ginger until fragrant.

2. Add seasoning. Cook over low heat until thick. Discard the spring onion, garlic and ginger.

Cooking Tips

★ Red vinasse is fermented glutinous rice in wine with red yeast rice. It is a natural and healthy condiment.

★ The maltose in the sauce tends to burn easily. Thus, make sure you stir the sauce continuously in the cooking process. Otherwise, it might burn and stick to the pot.

★ This sauce lasts in the fridge for 3 days.

鳳尾蝦

Prawn Toast

容易指數：★★★☆☆

◎ 材料

中蝦 6 隻
蝦仁 120 克
方包 1 塊（去硬邊，切成 6 件）
蛋汁 1 湯匙
生粉 1 湯匙

◎ 醃料

鹽 1/4 茶匙
胡椒粉少許

◎ 調味料

鹽 1/6 茶匙
生粉 1 茶匙
胡椒粉少許

◎ 做法

1. 中蝦去頭、去殼及去腸，留蝦尾，於背部輕剒一刀，用醃料拌勻。

2. 蝦仁用鹽及生粉擦勻，待半小時，沖洗，瀝乾水分，用乾布吸乾。

3. 蝦仁用刀拍爛，用刀背輕剁，下調味料順一方向拌至起膠，撻入碗內數次，冷藏片刻。

4. 方包塗上適量蝦膠。中蝦蘸上蛋汁及生粉，平放於蝦膠面，蝦尾向上。

5. 燒熱適量油，下方包炸至金黃色，趁熱品嘗。

◎ Ingredients

6 medium prawns
120 g shelled shrimp
1 sandwich bread (tough edge removed; cut into 6 pieces)
1 tbsp egg wash
1 tbsp caltrop starch

◎ Marinade

1/4 tsp salt
ground white pepper

◎ Seasoning

1/6 tsp salt
1 tsp caltrop starch
ground white pepper

◎ Method

1. Remove the head, shell and vein of the medium prawns. Reserve the tail. Cut a slit in the back. Mix with the marinade.

2. Rub the shelled shrimp with salt and caltrop starch. Leave for 1/2 hour. Rinse and drain. Wipe dry with a dry cloth.

3. Pound the shelled shrimp with a knife. Gently chop with the back of the knife. Add the seasoning. Stir in one direction until sticky. Throw into a bowl for several times. Refrigerate for a while.

4. Spread the sandwich bread with some minced shrimp. Dip the medium prawns in the egg wash and caltrop starch. Lay flat on the minced shrimp with the tail up.

5. Heat up some oil. Deep-fry the sandwich bread until golden. Serve warm.

鳳
尾
蝦

◯◯ 零失敗技巧 ◯◯
Successful cooking skills

打蝦膠很難，如何做得爽口？

去掉蝦殼後，毋須急於清洗，先用少許鹽及生粉拌勻，待一會才洗，用乾布徹底吸乾蝦肉，並順一方向攪至起膠，撻入碗內數次，冷藏一會，能做到爽口的蝦膠。

It is difficult to make minced shrimp. How to make it crisp?

It is no need to rinse the shelled shrimp immediately. Mix the shrimp with a little salt and caltrop starch. Leave them for a while and then rinse. Use a dry cloth to wipe dry thoroughly. Stir in one direction until it is sticky. Throw the paste into a bowl for several times and then refrigerate for a while. The minced shrimp will taste crisp.

若蝦膠一次用不完，如何處置？

放於密實盒內冷藏，可留待翌日使用。

How to deal with the remaining unused minced shrimp?

Put it in a sealed food box and refrigerate for use on the next day.

如何塗抹蝦膠？

將蝦膠平均塗抹方包上即可；但如想吃出豐富的蝦肉質感，可多塗一點蝦膠，啖啖蝦肉！

How do you spread the minced shrimp on the sandwich bread?

Just spread it evenly on the bread. If you prefer bouncy texture, spread on a thicker layer.

方包會吸收太多油分嗎？

首先將方包放入雪櫃冷藏至乾身，並在上碟前調大火候，將油分徹底迫出。

Will the sandwich bread absorb too much oil?

First refrigerate the sandwich bread until dry. Before serving, adjust to high heat and deep-fry the bread to press oil out thoroughly.

黑椒豬肉乾

Pork Jerky with Black Pepper

容易指數：★★★☆☆

材料

絞豬肉 500 克

調味料

生抽 2 湯匙
老抽半茶匙
魚露 1.5 湯匙
紹酒 1 湯匙
鹽半茶匙
糖 4 湯匙
蜂蜜 1.5 湯匙
黑椒碎半茶匙

做法

1. 絞豬肉先加入鹽、生抽循一方向攪至起膠，加入餘下的調味料再攪勻。

2. 將絞豬肉放在一張牛油紙上，豬肉面再覆上另一張牛油紙，用木棒擀薄約 2 厘米厚。

3. 移開豬肉面的牛油紙，將豬肉和底紙一起放入已預熱 160℃的焗爐內焗約 15 分鐘，再調至 170℃焗 20 分鐘，期間反轉 2 次，取出待涼，切片享用。

◯◯ Ingredient

500 g minced pork

◯◯ Seasoning

2 tbsp light soy sauce
1/2 tsp dark soy sauce
1.5 tbsp fish sauce
1 tbsp Shaoxing wine
1/2 tsp salt
4 tbsp sugar
1.5 tbsp honey
1/2 tsp ground black pepper

◯◯ Method

1 Add salt and light soy sauce to minced pork. Stir in one direction until sticky. Put in the remaining seasoning and mix well.

2 Place pork on a piece of baking paper. Put another piece of baking paper on top. Knead with a rolling pin until the pork is 2 cm thick.

3 Remove baking paper on top of the pork. Preheat oven at 160°C. Transfer the pork and the baking paper under it to the oven. Bake for 15 minutes. Turn to 170°C and bake for 20 minutes. Turn the pork upside down twice during baking. Remove and leave to cool. Slice and serve.

◯◯ 零失敗技巧 ◯◯
Successful cooking skills

自製豬肉乾，怎樣才美味？
將絞豬肉醃一夜才烘焙，味道更佳。
What is the secret trick to tasty pork jerky?
Marinate the pork overnight before baking it. It will taste better that way.

要注意哪個步驟？
必須用牛油紙覆蓋豬肉面才擀薄，否則肉碎黏滿木棒；而且，豬肉擀薄後更香口。
Is there any step that needs my special attention?
Cover the pork with baking paper before rolling it out with a rolling pin. Otherwise, the pork would stick to the rolling pin. The pork jerky also tastes better if rolled thinner.

酥炸棒棒雞

Deep-fried Chicken Drumsticks

容易指數：★★★☆☆

材料

雞鎚 8 隻

醃料

鹽 1/3 茶匙
雞粉 1/3 茶匙
糖 1/8 茶匙
胡椒粉少許

脆漿料

麵粉 6 湯匙
發粉半茶匙
生粉 1.5 湯匙
清水 7 湯匙
油 1 湯匙（後下）

甜辣醬（拌勻）

蛋黃醬 1 湯匙
茄汁 3/4 湯匙
是拉差辣椒醬 1 茶匙
煉奶 1 茶匙
凍開水 1 茶匙

做法

1. 雞鎚解凍，洗淨，用利剪沿骨剪開雞肉，將雞肉褪至雞鎚頂端，雞肉向外，拌入醃料醃半小時。（參考 p.10）

2. 脆漿料拌勻待半小時，再加入生油混和，備用。

3. 雞鎚肉蘸上脆漿，放入熱油炸至金黃色及熟透，上碟，蘸甜辣醬伴食。

Ingredients

8 drumsticks

Marinade

1/3 tsp salt
1/3 tsp chicken bouillon powder
1/8 tsp sugar
ground white pepper

Batter ingredients

6 tbsp flour
1/2 tsp baking powder
1.5 tbsp caltrop starch
7 tbsp water
1 tbsp oil (for later use)

Sweet chilli sauce (mixed well)

1 tbsp mayonnaise
3/4 tbsp ketchup
1 tsp Sriracha chilli sauce
1 tsp condensed milk
1 tsp cold drinking water

Method

1. Defrost the drumsticks. Rinse. Cut off the meat along the bone with a pair of sharp scissors. Withdraw the meat to the top of the drumsticks with the meat on the outside. Mix with the marinade and rest for 1/2 hour. (refer to p.10)

2. Mix the batter ingredients and rest for 1/2 hour. Mix in the oil. Set aside.

3. Dip the drumsticks in the batter. Deep-fry in hot oil until golden and thoroughly cooked. Put on the plate. Serve with the sweet chilli sauce.

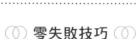

◎ 零失敗技巧 ◎
Successful cooking skills

褪雞鎚肉有何技巧？

試試由細骨向粗骨方向剪去，完成後雞鎚直立，賣相吸引！

How to withdraw the meat from the drumstick skilfully?

Try scissoring along the bone from bottom to top (thin to thick). When finished, the drumstick is upright and attractive!

用新鮮雞鎚炸透，會浪費嗎？

當然不會！更可品嘗鮮嫩的肉汁，美味啊！

Is it a waste to deep-fry fresh drumsticks?

Of course not! It is fresh, succulent and delicious!

拌脆漿料時，生油為何最後拌入？有何作用？

若太早拌入生油，會妨礙麵粉及發粉的發酵效果，令脆漿酥脆不足！

Why add oil at last when mixing the batter ingredients?

Mixing in the oil too early will make the proving of flour and baking powder less effective. The batter will become less crisp after deep-frying!

甜辣醬包含蛋黃醬、茄汁、是拉差辣椒醬及煉奶，味道如何？

蛋黃醬香滑；茄汁甜酸；是拉差醬辣香；煉奶香滑，出奇配合，百味紛陳，令人一試難忘！

Sweet chilli sauce comprises mayonnaise, ketchup, Sriracha chilli sauce and condensed milk. It has a weird flavour, isn't it?

The mixed flavour – smooth mayonnaise, sweet and sour ketchup, hot and spicy Sriracha chilli sauce, fragrant condensed milk – is complicated, but they match dramatically and make you unforgettable in just one bite!

香茅雞肉串

Lemongrass Chicken Skewers

容易指數：★★★☆☆

◎ 材料
雞肉 250 克
紅咖喱醬 1.5 茶匙
麥片 3 湯匙（用水浸軟）
水 3 湯匙
檸檬葉 1 片（去梗、切碎）
香茅 5 枝（斜切成段）
生粉 2 湯匙

◎ 調味料
鹽半茶匙
魚露 1 湯匙
糖半茶匙
胡椒粉少許

做法

1. 雞肉攪碎或用刀剁碎成免治雞肉，加入鹽及魚露，順一方向攪至起膠，再加入糖及胡椒粉攪拌均勻。

2. 拌入麥片、紅咖喱醬及檸檬葉碎，放入雪櫃冷藏片刻。

3. 用保鮮紙放於手掌上，舀 2 湯匙雞茸料於於保鮮紙，略壓平，將香茅斜口一端包入雞茸內，用保鮮包捏好，輕輕撲上生粉。

4. 燒熱少許油，用半煎炸方式將香茅雞肉串煎成金黃色，伴泰式甜酸醬享用。

Ingredients

250 g chicken meat
1.5 tsp red curry paste
3 tbsp oats (soaked in water to soften)
3 tbsp water
1 lime leaf (removed the stem; chop up)
5 stalks lemongrass (cut diagonally into sections)
2 tbsp cornflour

Seasoning

1/2 tsp salt
1 tbsp fish sauce
1/2 tsp sugar
ground white pepper

香茅雞肉串

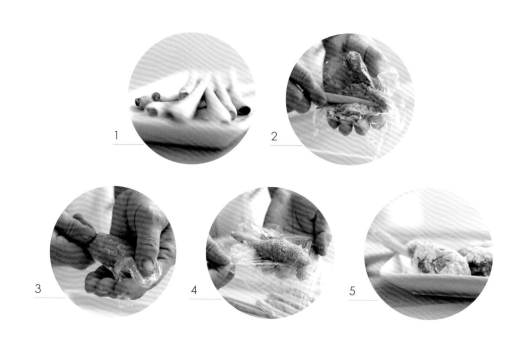

◎◎ Method

1. Mince the chicken meat with a food processor or a knife. Mix the minced chicken with the salt and fish sauce. Stir in one direction until sticky. Add the sugar and ground white pepper. Mix well.

2. Stir in the oats, red curry paste and lime leaf. Refrigerate for a while.

3. Place cling film on the palm. Put 2 tbsp of the chicken paste onto the cling film. Slightly flatten. Wrap one end of the lemongrass in the chicken paste. Shape with the cling film. Lightly coat with the caltrop starch.

4. Heat up a little oil. Fry the lemongrass chicken skewers until golden. Serve with Thai sweet chilli sauce.

◎◎ 零失敗技巧 ◎◎
Successful cooking skills

如何弄出惹味的雞肉茸？

雞肉解凍後，徹底吸乾水分，剁爛，拌至起膠，再與調味料混和，必定美味可口！

Is there any way to make the minced chicken extra-tasty?

After thawing the chicken, wipe it dry completely. Then finely chop it and stir until it turns sticky. Add seasoning and mix well. It is guaranteed to taste good that way.

雞茸餡加入了麥片，口感如何？

拌入麥片後，雞茸更彈牙好吃，而且增加食物水溶性纖維，一舉兩得！

Why add oats in the chicken paste?

By adding oats, the chicken paste is more spongy and delicious, and contains more soluble dietary fiber.

如何有效提升香茅的味道？

香茅別太早斜切成段，以免香茅容易乾透，香氣過早揮發！

How to enhance the flavour of lemongrass?

Cut the lemongrass just before cooking to avoid it drying and releasing the aroma too early.

馬介休蟹肉薯球

Deep-fried Potato Balls with Bacalhau and Crabmeat

容易指數：★★★☆☆

◎ 材料

鮮蟹肉 2 兩（參考 p.6）
馬介休 1 兩
馬鈴薯 10 兩
洋葱 1 兩

◎ 調味料

鹽 1/4 茶匙
生抽 2 茶匙
胡椒粉少許
生粉 3 湯匙

◎ 做法

1. 馬介休抹淨，去骨、去皮，切幼粒。

2. 馬鈴薯去皮，磨成幼絲；洋葱切碎。

3. 馬介休、薯絲、蟹肉、洋葱及調味料拌勻，搓成薯球（約 12 個）。

4. 燒熱油，下薯球炸至金黃香脆，上碟品嘗。

Ingredients

75 g fresh crabmeat (refer to p.6)
38 g Bacalhau (salted codfish)
375 g potatoes
38 g onion

Seasoning

1/4 tsp salt
2 tsp light soy sauce
ground white pepper
3 tbsp caltrop starch

Method

1. Wipe Bacalhau. Remove the bone and skin. Dice finely.

2. Peel potatoes and grate into fine shreds. Chop the onion.

3. Mix Bacalhau, potato shreds, crabmeat, onion and seasoning. Knead into about 12 potato balls.

4. Heat oil in wok. Deep-fry the potato balls until golden brown and crispy. Serve.

零失敗技巧
Successful cooking skills

甚麼是馬介休？哪裏有售？
馬介休是葡萄牙之醃製食材，是鱈魚用鹽醃成，於售賣澳門乾貨店舖有售。

What is Bacalhau? Where to buy it?

Bacalhau (in Portuguese) is a kind of pickled food from Portuguese. It is salted codfish and can be bought from grocery stores selling Macau products.

可以用薯茸炮製嗎？
將馬鈴薯刨成絲，咬入口，絲絲的質感令口感非常豐富！

May I use the potato puree?

I suggest to use the potato shreds. It enhance the texture of the dish.

馬介休之鹹味會否掩蓋蟹肉之鮮味？
絕對不會，馬介休的鹹味較淡，使用 1 兩的份量不會掩蓋蟹肉之鮮味。

Would the salty taste of Bacalhau cover the fresh taste of crabmeat?

Bacalhau has a light salty taste only and 38 g of it would not cover the fresh taste of crabmeat.

金盞海鮮燴

Braised Seafood and Zucchini in Crispy Cups

容易指數：★★☆☆☆

◎ **材料**
中蝦 8 隻
急凍帶子 4 隻
鮮蟹肉 3 兩（參考 p.6）
魚柳 3 兩
翠玉瓜 4 兩
方型大雲吞皮 8 張
蒜茸半湯匙
甘筍 1 條（直徑不超過 5 厘米）

◎ **醃料（1）**
鹽 1/4 茶匙
胡椒粉少許

◎ **醃料（2）**
鹽 1/3 茶匙
胡椒粉及麻油各少許
蛋白 1 茶匙
生粉半湯匙

調味料
鹽半茶匙
水適量

 獻汁（拌勻）
水 1/4 杯
鹽及糖各 1/4 茶匙
生抽半茶匙
麻油少許
生粉半茶匙

 做法
1. 甘筍洗淨及抹乾水分。
2. 燒熱油一杯（見出現小泡沫即可），在甘筍較粗一端包上雲吞皮，放入油內炸至金黃色及香脆的金盞，脫開甘筍，放於碟上。
3. 中蝦去殼、去腸，洗淨，抹乾後與醃料（1）拌勻。
4. 帶子解凍，每隻切為兩份；魚柳切粗粒；帶子、魚柳及醃料（2）拌勻醃 10 分鐘。
5. 煮滾適量清水，下帶子浸 1 分鐘，盛起。
6. 翠玉瓜洗淨，切粗粒，與調味料炒熟，盛起。
7. 熱鑊下油，加入中蝦、魚柳及帶子，下蒜茸拌炒至香氣四溢，加入蟹肉、獻汁及翠玉瓜炒勻，分放入金盞內享用。

 Ingredients
8 medium-sized prawns
4 frozen scallops
113 g fresh crabmeat (refer to p.6)
113 g fish fillet
150 g zucchini
8 square large wanton wrappers
1/2 tbsp finely chopped garlic
1 carrot (diameter not exceeds 5 cm)

 Marinade (1)
1/4 tsp salt
ground white pepper

 Marinade (2)
1/3 tsp salt
ground white pepper
sesame oil
1 tsp egg white
1/2 tbsp caltrop starch

 Seasoning
1/2 tsp salt
water

 Thickening glaze (mixed well)
1/4 cup water
1/4 tsp salt
1/4 tsp sugar
1/2 tsp light soy sauce
sesame oil
1/2 tsp caltrop starch

◎◎ Method

1. Rinse and wipe dry carrot.

2. Heat a cup of oil in wok (until little bubbles appear). Wrap wanton wrapper at the thicker side of carrot. Deep-fry in oil until golden brown. This becomes crispy golden cup. Remove the carrot and put the cup in a plate.

3. Shell and devein prawns. Rinse and wipe dry. Mix well with marinade (1).

4. Defrost scallops and each cut into 2 pieces. Cut fish fillet into thick dices. Mix scallops and fish fillet with marinade (2) and set aside for 10 minutes.

5. Bring water to the boil. Add scallops and soak for 1 minute. Drain.

6. Rinse zucchini and cut into thick dices. Stir-fry with the seasoning until done and set aside.

7. Add oil into a hot wok. Put in prawns, fish fillet and scallops. Add finely chopped garlic and stir-fry until fragrant. Pour in crabmeat, thickening glaze and zucchini. Stir-fry well and put into the crispy cups. Serve.

金盞海鮮燴

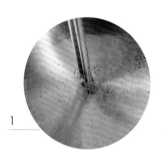
1

2

3

◎◎ 零失敗技巧 ◎◎
Successful cooking skills

炸金盞容易處理嗎？

雲吞皮輕輕地包在甘筍，放入熱油內，雲吞皮迅速炸至香脆及脫開，處理簡易，緊記油溫不要過熱。

Is it easy to deep-fry crispy cup?

Wrap wanton wrapper around the carrot and deep-fry in hot oil. The wrapper becomes crispy and demould quickly. The steps are simple but remember that the oil cannot be too hot.

黃金糯米釀雞翼

Chicken Wings Stuffed with Salted Egg Yolk and Glutinous Rice

容易指數：★☆☆☆☆

◎ 材料

急凍雞全翼 4 隻
糯米 2 兩
蝦米 1 兩
鹹蛋黃 2 個
生粉適量
牙籤 4 支

◎ 醃料

味椒鹽半茶匙
老抽半茶匙

◎ 調味料

鹽 1/8 茶匙
蠔油 1 茶匙
糖 1/4 茶匙
胡椒粉及麻油各少許

◎ 做法

1. 糯米洗淨，用水浸 4 小時，隔去水分，蒸 20 分鐘，放入鹹蛋黃再蒸 10 分鐘。鹹蛋黃切粒；蝦米浸軟，備用。

2. 雞全翼解凍，切去雞鎚部分，於骨位剪開筋部，沿骨位剪入，慢慢褪出雞翼骨。（參考 p.11）

3. 將去骨雞翼洗淨，吸乾水分，用醃料拌勻醃約半小時。

4. 熱鑊下油，下蝦米及鹹蛋黃爆香，加入糯米及調味料炒勻，待涼。

5. 將步驟（4）的材料釀入雞翼內（約八成滿），用牙籤串起。

6. 雞翼撲上薄薄的生粉，放入熱油炸至金黃色及熟透享用。

黃金糯米釀雞翼

◎ 零失敗技巧 ◎
Successful cooking skills

難以釀入餡料，為甚麼？
我建議選購體型較大的雞全翼，容易釀入餡料。
Why is it hard to stuff the chicken wings?
I suggest using large whole chicken wings for easy stuffing.

若餡料釀得太滿，後果如何？
入鑊炸時，雞翼會收縮，餡料受熱後容易爆瀉，影響賣相及口感，故餡料別釀得太多太滿。
What will happen if stuffed with too much filling?
The chicken wing will contract in the process of deep-frying. Too much filling when heated will make the filling burst, giving a poor presentation and taste. Do not stuff more than enough.

Ingredients

4 frozen whole chicken wings
75 g glutinous rice
38 g dried shrimp
2 salted egg yolks
caltrop starch
4 toothpicks

Marinade

1/2 tsp flavoured pepper salt (Aji Shio Kosho)
1/2 tsp dark soy sauce

Seasoning

1/8 tsp salt
1 tsp oyster sauce
1/4 tsp sugar
ground white pepper
sesame oil

Method

1. Rinse the glutinous rice. Soak in water for 4 hours. Drain. Steam for 20 minutes. Put in the salted egg yolks and steam for 10 minutes. Dice the salted egg yolks. Soak the dried shrimp until soft. Set aside.

2. Defrost the chicken wings. Cut off the drumsticks. Cut the tendon off the bone with a pair of scissors. Cut along the bone. Remove the bone slowly. (refer to p.11)

3. Rinse the boned chicken wings. Wipe dry. Mix with the marinade and rest for about 1/2 hour.

4. Add oil in a heated wok. Stir-fry the dried shrimp and salted egg yolk until fragrant. Add the glutinous rice and seasoning. Stir-fry and mix well. Let it to cool.

5. Stuff the chicken wings (in about 80% full) with the filling from step 4. Skewer with the toothpicks.

6. Thinly coat the chicken wings with the caltrop starch. Deep-fry in hot oil until golden and fully done. Serve hot.

1

2

煎三文鹹魚肉餅

Pan-fried Pork Patties with
Flaked Salted Salmon and Dried Scallops

容易指數：★★★☆☆

材料

免治豬肉 6 兩
瑤柱三文魚醬 3 茶匙（參考 p.12）
雞蛋半個
粟粉適量

調味料

生抽半湯匙
糖 1/4 茶匙
胡椒粉少許
粟粉 2 茶匙

做法

1. 免治豬肉拌入雞蛋及調味料，用筷子攪拌至起膠，下瑤柱三文魚醬拌勻。

2. 將肉醬弄成小圓餅，沾上乾粟粉，放入油鑊用中小火煎至兩面金黃全熟，趁熱享用。

Ingredients

225 g ground pork
3 tsp flaked salted salmon with dried scallops (refer to p.12)
1/2 egg
cornflour

Seasoning

1/2 tbsp light soy sauce
1/4 tsp sugar
ground white pepper
2 tsp cornflour

Method

1. Stir the egg and seasoning into the pork. Stir with chopsticks until sticky. Add the flaked salted salmon with dried scallops. Stir well.

2. Shape the pork mixture into round patties. Coat them lightly in cornflour. Fry in a little oil over medium-low heat until both sides golden and fully done. Serve hot.

◎◎ 零失敗技巧 ◎◎
Successful cooking skills

如何將豬肉攪至起膠？
建議選用深且大的碗，快速順一方向攪拌，容易掌握。
How can you make the pork patties sticky and chewy?
Put the pork into a deep big bowl. Stir in one direction with chopsticks until sticky.

為何用筷子攪拌？
筷子比用湯匙更能快速令豬肉起膠。
Why do you stir the pork with chopsticks?
The pork tends to get sticky more quickly when you stir it with chopsticks rather than a spoon.

豬肉起膠的狀態是怎樣？
見豬肉及配料黏稠、不鬆散，按下去有彈力即可。
How would I know the pork stirring until sticky?
You will see the pork turn to be sticky and not too loosen. Feel elastic if pressing the pork.

可以買現成的免治豬肉嗎？
基本上可以的，但若想吃時口感更豐富，建議自行用刀剁成免治豬肉，很方便！
Can I use pork that has been pre-ground?
Basically, you can. But it tastes best if you chop a whole piece of pork yourself. Don't be intimidated. It's less complicated than you think.

腐皮石榴雞

Steamed Chicken Parcels in Tofu Skin

容易指數：★★☆☆☆

◎ 材料

鮮腐皮 1 塊
雞肉 3 兩
沙葛 3 兩
乾冬菇 3 朵
蛋白 1 個
韭菜 8 條
芫茜碎適量

◎ 調味料（調勻）

蠔油半湯匙
糖 1/4 茶匙
胡椒粉少許
粟粉半茶匙

◎ 獻汁（調勻）

鹽 1/6 茶匙
粟粉 1 茶匙
麻油 2 茶匙，水半杯

◎ 做法

1. 鮮腐皮剪去硬邊，用乾淨濕毛巾略抹，剪成 8 方塊（13 厘米 x 13 厘米）。

2. 韭菜洗淨，用滾水灼軟，盛起備用；冬菇去蒂，用水浸軟，切粒；雞肉洗淨，切粒；沙葛去皮，洗淨，切粒。

3. 燒熱鑊，下油 1 湯匙，放入雞肉、冬菇及沙葛炒片刻，下調味料炒至雞肉熟透，最後加入芫茜碎拌勻，待涼成餡料。

4. 腐皮鋪平，放入適量餡料，用韭菜紮緊，排於碟上，隔水大火蒸 8 分鐘，傾去汁液。

5. 煮滾獻汁，下蛋白拌勻，澆在石榴雞上即可享用。

◎ 零失敗技巧 ◎
Successful cooking skills

如沒有沙葛可用甚麼代替？
可嘗試用馬蹄代替，同樣有爽脆的口感，但建議購買帶皮的馬蹄，回家自行去掉外皮。

If yam bean is not available at the market, what can we use?

Having a crunchy texture, water chestnuts can be used instead. It is better to buy the whole water chestnuts and skin them at home.

如何用韭菜紮緊腐皮？
勿釀入太多餡料，慢慢用韭菜綑好，耐心地處理必成。

How to tie tofu skin tightly with Chinese chive?

With patience, you can make it. Remember not to stuff with too much filling. Then tie the tofu skin slowly with the Chinese chive.

Ingredients

1 piece fresh tofu skin
113 g chicken meat
113 g yam bean
3 dried black mushrooms
1 egg white
8 strips Chinese chives
chopped coriander

Seasoning (mixed well)

1/2 tbsp oyster sauce
1/4 tsp sugar
ground white pepper
1/2 tsp cornflour

Thickening glaze (mixed well)

1/6 tsp salt
1 tsp cornflour
2 tsp sesame oil
1/2 cup water

Method

1. Cut away the hard edge of the tofu skin. Slightly wipe it with a clean damp towel. Cut into 8 squares (13cm x 13cm).

2. Rinse the Chinese chives. Blanch until soft. Set aside. Remove the stalks of the dried black mushrooms. Soak in water until soft and dice. Rinse and dice the chicken meat. Remove the skin of yam bean. Rinse and dice.

3. Heat a wok. Put in 1 tbsp of oil. Stir-fry the chicken meat, mushrooms and yam bean for a moment. Add the seasoning and stir-fry until the chicken meat is fully done. Stir in the coriander at last to become the filling. Set aside to cool.

4. Lay flat the tofu skin. Put in some of the filling. Tie with the Chinese chives tightly as parcels. Arrange them on a plate. Steam over high heat for 8 minutes. Pour away the steaming sauce.

5. Bring the thickening glaze to the boil. Stir in the egg white. Drizzle on top of the chicken parcels. Serve.

蝦子竹笙百花雞卷

Chicken and Shrimp Puree
Stuffed in Zhu Sheng Rolls with Shrimp Roe

 容易指數：★★☆☆☆

◎ 材料
竹笙 10 條
即食蝦子 1 茶匙
中蝦 4 兩
雞柳 2 兩
菜心 4 兩
清雞湯 2/3 杯（150 毫升）

◎ 飛水材料
紹酒半湯匙
薑 1 片
葱 1 條

◎ 醃料
鹽 1/4 茶匙
生抽 1 茶匙
胡椒粉及麻油各少許
生粉半茶匙

44

調味料

油半湯匙

鹽 1 茶匙

 獻汁

清雞湯 1/3 杯（100 毫升）

鹽及糖各 1/4 茶匙

生抽半茶匙

麻油及胡椒粉各少許

生粉半茶匙

 做法

1. 竹笙浸水約 2 小時（期間換水 1 至 2 次），剪去一端，修剪成 6 厘米長度。

2. 竹笙放入滾水內，放入飛水材料煮 5 分鐘，盛起，沖淨，吸乾水分，備用。

3. 中蝦去殼、去腸，洗淨，用刀拍扁，剁成蝦膠。

4. 雞柳去筋，剁成雞茸，與蝦膠及醃料順一方向拌至帶黏性，分成 10 等份。

5. 將步驟（4）的百花雞茸餡料放入擠袋，慢慢擠入竹笙至半飽滿，剪去竹笙末端，用手輕輕按壓餡料。

6. 竹笙卷放碟上，傾入清雞湯，隔水蒸 6 分鐘，傾出蒸汁。

7. 菜心修剪成菜遠，放入已下調味料的滾水內，灼熟，伴於竹笙卷旁。

8. 燒滾獻汁，澆在竹笙百花雞卷上，最後灑上蝦子即成。

 Ingredients

10 pieces Zhu Sheng

1 tsp cooked shrimp roe

150 g medium prawns

75 g chicken fillet

150 g Chinese flowering cabbage

2/3 cup chicken stock (150 ml)

 Ingredients for scalding

1/2 tbsp Shaoxing wine

1 slice ginger

1 sprig spring onion

 Marinade

1/4 tsp salt

1 tsp light soy sauce

ground white pepper

sesame oil

1/2 tsp caltrop starch

 Seasoning

1/2 tbsp oil

1 tsp salt

 Thickening glaze

1/3 cup chicken stock (100 ml)

1/4 tsp salt

1/4 tsp sugar

1/2 tsp light soy sauce

sesame oil

ground white pepper

1/2 tsp caltrop starch

Method

1. Soak the Zhu Sheng in water for about 2 hours (change water for 1 to 2 times). Cut away one end. Trim into 6 cm long.

2. Put the Zhu Sheng in boiling water. Add the ingredients for scalding. Scald for 5 minutes. Rinse and wipe dry. Set aside.

3. Shell and devein the medium prawns. Rinse. Flatten with a knife. Finely chop into paste.

4. Cut away the veins of the chicken fillet. Chop into chicken puree. Mix with the minced prawn and marinade. Stir in one direction until sticky. Divide evenly into 10 portions.

5. Put the prawn and chicken filling from step 4 into a piping bag. Pipe into the Zhu Sheng slowly until it is half full. Cut away the end of the Zhu Sheng. Press the filling lightly by hand.

6. Put the Zhu Sheng rolls on a plate. Pour in the chicken stock. Steam over water for 6 minutes. Remove the steaming sauce.

7. Trim the Chinese flowering cabbage. Use only the inner parts. Put in boiling water added with the seasoning. Blanch until done. Put on the side of the rolls.

8. Bring the thickening glaze to the boil. Pour over the rolls. Sprinkle with the shrimp roe. Serve.

蝦子竹笙百花雞卷

零失敗技巧
Successful cooking skills

挑選竹笙有何秘訣？

挑選帶淡黃色澤的野生竹笙，沒經人工漂白，而且野生竹笙比人工培埴的更爽口，向店員詢問清楚。

How to select Zhu Sheng?

Choose those that are light yellow in colour and not bleached. Wild Zhu Sheng have a crunchier texture compared with the cultivated. Consult the shop staff for the best ones.

如何剁成嫩滑的雞茸？

必須用刀去掉雞柳上白色的筋，而且將雞肉徹底剁幼，雞茸必然嫩滑好吃。

How to make the chicken puree smooth?

It can be done by cutting away the white veins of the chicken fillet with a knife and chopping finely.

魚湯鮮菇浸魚腐

Hon-shimeji Mushrooms and Tofu Fish Puffs in Fish Stock

容易指數：★★★☆☆

◎◎ 魚腐材料
新鮮鯪魚脊 2 條 （約 10 兩）
嫩滑板豆腐 1 塊
雞蛋 2 個
粟粉 3 湯匙

◎◎ 配料
本菇 1 包
薑 4 片

◎◎ 調味料
鹽 1 茶匙
糖半茶匙
胡椒粉適量

◎◎ 魚腐做法
1. 鯪魚脊洗淨，起肉，去皮（鯪魚骨及魚皮留用），剁成茸，加入調味料拌至起膠。
2. 魚肉、豆腐、雞蛋及粟粉順一方向拌成豆腐魚漿，舀起魚肉漿，放入滾油內炸至金黃色，隔油。

◎◎ 做法
1. 本菇切去尾部，洗淨，備用。
2. 鯪魚骨及魚皮洗淨，加入薑片，放入油鑊煎至微黃色，隔去多餘油分，傾入滾水 4 杯，用中火煲 30 分鐘，盛起鯪魚骨及鯪魚皮，加入適量魚腐煮滾，最後放入本菇煮滾即成。

◎◎ Ingredients for tofu fish puffs
2 backbones of dace with meat (about 380 g)
1 fresh firm tofu
2 eggs
3 tbsp cornflour

◎◎ Condiments
1 packet Hon-shimeji mushrooms
4 slices ginger

◎◎ Seasoning
1 tsp salt
1/2 tsp sugar
ground white pepper

◎◎ Method for tofu fish puffs
1. Rinse, skin and free the meat from the backbones of dace (reserve the bones and skin). Finely chop the meat. Add the seasoning and stir until sticky.
2. Stir the meat, tofu, eggs and cornflour in one direction to become fish puff paste. Spoon the paste into hot oil. Deep-fry until golden brown. Drain.

◎◎ Assembly
1. Cut away the root of the Hon-shimeji mushrooms. Rinse and set aside.
2. Rinse the fish bones and skin. Fry with the ginger in a wok until light brown. Remove excessive oil. Pour in 4 cups of hot water. Cook over medium heat for 30 minutes. Remove the bones and skin. Add some tofu fish puffs and bring to the boil. Put in the mushrooms at last and bring to the boil again. Serve.

1　　2

3　　4　　5

◯◯ 零失敗技巧 ◯◯
Successful cooking skills

魚腐如何弄得彈牙好吃？

拌豆腐及魚肉等材料時，必須快速攪拌，令魚肉起膠，肉質彈牙。

How to make tofu fish puffs springy and tasty?

Stir tofu, fish flesh and other ingredients swiftly to make it gluey and springy.

鯪魚脊容易購買嗎？價錢昂貴？

街市的淡水魚檔經常有售，價錢便宜。用鯪魚脊熬湯，湯水美味且不油膩。

Is the meat on the backbone of dace easily available? Is it expensive?

It is economical and easily available at freshwater fish stalls in the market. The soup stewed with the backbone is flavourful and not greasy.

用魚皮及魚骨煲湯的效果如何？

鯪魚皮及魚骨用油略煎，注入滾水只需熬 30 分鐘，可弄成奶白鮮滑的魚湯。

How about making the soup with the skin and the bone?

Slightly fry the fish skin and bone. Pour in hot water and cook for 30 minutes. Then you will enjoy a smooth and milky soup.

炸芋茸腐皮卷

Deep Fried Taro Rolls in Tofu Skin

容易指數：★★★☆☆

材料

鮮腐皮 1 張
芋頭半個
澄麵 2 湯匙
雞蛋 1 個（拂勻）

調味料

五香粉 1/4 茶匙
鹽 1/4 茶匙

甜酸汁

米醋 4 湯匙
茄汁 3 湯匙
片糖半塊（舂碎）
鹽 1/8 茶匙

做法

1. 芋頭去皮，洗淨，切粗粒，隔水用大火蒸熟，趁熱用刀壓成芋茸，加入澄麵及調味料，拌成芋泥。

2. 鮮腐皮剪去硬邊，用乾淨的濕毛巾略抹，抹上薄薄的芋泥（塗於腐皮2/3 位置），捲成長條形，用蛋液黏緊。

3. 甜酸汁用慢火煮至片糖溶化，盛起備用。

4. 燒滾油，放入芋茸卷，轉慢火炸至金黃色，盛起，瀝乾油分，切塊，伴甜酸汁蘸吃。

Ingredients

1 piece fresh tofu skin
1/2 taro
2 tbsp wheat starch (Tang flour)
1 egg (whisked)

Seasoning

1/4 tsp five-spice powder
1/4 tsp salt

Sweet and sour sauce

4 tbsp rice vinegar
3 tbsp ketchup
1/2 slab brown sugar (crushed)
1/8 tsp salt

Method

1. Skin the taro. Rinse and coarsely dice. Steam over high heat until done. Mash with a knife while hot. Add the wheat starch and the seasoning. Mix well as taro puree.

2. Cut away the tough edge of the tofu skin. Slightly wipe with a clean damp towel. Thinly spread the taro puree on 2/3 parts of the skin. Roll into a strip. Seal with the egg wash.

3. Cook the sweet and sour sauce over low heat until the brown sugar slab melts. Dish up and set aside.

4. Put the roll in hot oil. Then turn to low heat and deep-fry until golden brown. Dish up and drain. Cut into pieces. Serve with the sweet and sour sauce.

◯◯ 零失敗技巧 ◯◯
Successful cooking skills

如何避免芋泥溢出？
於芋茸卷兩端抹上生粉，以免芋泥容易溢出。
How to prevent taro puree from coming out?
Spread cornflour on both ends of the roll to seal in the taro puree.

鮮腐皮如何儲存？
先用保鮮袋盛好，再用報紙包裹，放於蔬菜冷藏格，可保存約兩星期。
How to keep fresh tofu skin?
Put the skin in a food bag. Then wrap it in a newspaper and keep it in the vegetable compartment of a refrigerator. It can last for about 2 weeks.

加入澄麵有何作用？
與芋茸拌勻後，令芋茸更黏稠。
What is the purpose of adding wheat starch?
It is to make the taro puree thick and gluey after mixing.

芋茸卷切塊後炸透，可以嗎？
建議將整條芋茸卷下油鍋炸透，以免芋茸餡乾硬，欠口感。
Is it possible to cut the roll into pieces and then deep-fry?
To avoid the filling turning hard and dry, it is better to deep-fry the whole roll first.

 # XO 醬帶子炒鮮露筍

Stir-fried Scallops and Asparaguses in XO Sauce

容易指數：★★★★☆

◐ 材料

急凍帶子 6 兩
鮮露筍 6 兩
鮮百合 1 球
小粟米 8 條
XO 醬 2 茶匙（參考 p.14）
紹酒半湯匙
蒜肉 3 粒

◐ 醃料

胡椒粉少許
粟粉 2 茶匙

◐ 調味料

蠔油 2/3 湯匙
生抽 2 茶匙
糖半茶匙
粟粉 1 茶匙
水 2 湯匙

◐ 做法

1. 帶子解凍，洗淨，抹乾水分，下醃料拌勻，飛水，瀝乾水分。

2. 鮮百合切去焦黃部分，撕成瓣狀，洗淨，用水浸過面備用。

3. 鮮露筍只取翠嫩部分，洗淨，切斜段；小粟米切段，洗淨；鮮露筍及小粟米飛水，瀝乾水分。

4. 燒熱鑊下油 2 湯匙，下蒜肉及 XO 醬炒香，加入帶子，灒酒炒勻，下鮮露筍、小粟米及鮮百合炒勻，最後下調味料（傾入時不斷翻炒），再炒片刻即成。

◐ Ingredients

225 g frozen scallops
225 g asparaguses
1 fresh lily bulb
8 baby sweet corns
2 tsp XO sauce (refer to p.14)
1/2 tbsp Shaoxing wine
3 cloves skinned garlic

◐ Marinade

ground white pepper
2 tsp cornflour

◐ Seasoning

2/3 tbsp oyster sauce
2 tsp light soy sauce
1/2 tsp sugar
1 tsp cornflour
2 tbsp water

◐ Method

1. Thaw the scallops. Rinse well and wipe dry. Add marinade and mix well. Blanch them in boiling water. Drain.

2. Cut off the dark yellow base of the lily bulb. Tear into scales. Rinse well. Soak it completely in water. Set aside.

3. Cut off the fibrous end of the asparaguses. Rinse. Cut into short lengths at an angle. Cut the baby sweet corns into short lengths. Rinse. Blanch the asparaguses and baby sweet corns. Drain.

4. Heat a wok and add 2 tbsp of oil. Stir fry garlic and XO sauce until fragrant. Put in the scallops. Sizzle with wine and stir fry well. Put in the asparaguses, corns and lily bulbs. Toss well. Add seasoning at last while stirring continuously. Stir-fry for a while. Serve.

XO醬帶子炒鮮露筍

◎ 零失敗技巧 ◎
Successful cooking skills

急凍帶子如何解凍？

從雪櫃的冰箱格移放至下層，慢慢解凍即可，若時間急趕，可用少許水浸至軟身。

How do you thaw the frozen scallops?

Transfer them from the freezer to a lower shelf of the refrigerator. Leave it them to thaw slowly until soft. If you're pressed for time, you may soak them in a little lightly salted water until soft.

為何鮮百合用水浸透？

鮮百合容易氧化，浸於水內能保持潔白色澤。

Why do you soak the fresh lily bulb in water?

Fresh lily bulb gets oxidized easily when exposed to air. Soaking them in water helps keep them white.

帶子如何炒至肉質爽彈？

帶子先飛水能縮短炒煮時間，以免肉質容易變韌。

What are the tricks to crunchy and resilient scallops?

You should blanch them first. It helps shorten the stir-frying time and thus, the scallops are less likely to get overcooked.

燒腩仔

Roasted Pork Belly

容易指數：★☆☆☆☆

◎ 材料
急凍五花腩 1.5 公斤
粗鹽 300 克

◎ 醃料
鹽 1 湯匙
五香粉及沙薑粉各半茶匙
玫瑰露酒 1 湯匙

◎ 工具
錫紙 1 張
棉繩 1 條

◎ 做法

1. 急凍五花腩放於雪櫃下層自然解凍，放入沸水煮約 10 分鐘，盛起，過冷河，抹乾。

2. 用針插或叉子在五花腩皮戳入，令整塊皮層戳滿小孔。

3. 反轉腩肉，用刀剁入兩刀，約 1.5 厘米深，塗上玫瑰露酒，再擦勻五香粉及沙薑粉，醃約 3 至 4 小時。

4. 錫紙對摺，摺成長條狀，沿五花腩四周圍起，高於五花腩約 2 厘米，用棉繩紮好。

5. 將粗鹽均勻地鋪於五花腩面，約 1 厘米厚。

6. 將整個五花腩放於焗盤，放入預熱焗爐以 200℃ 焗約 45 分鐘，取出。

7. 取走面層的粗鹽，再放入焗爐以上火焗約 15 分鐘，至五花腩皮層呈金黃色及香脆，待涼，斬件享用。

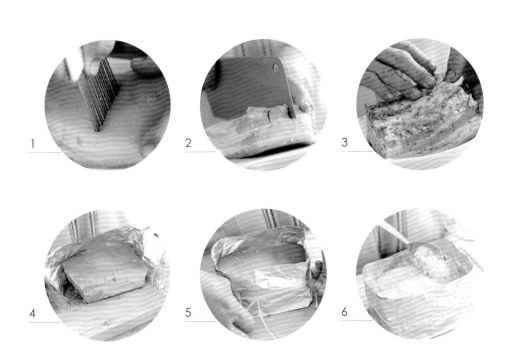

1   2   3

4   5   6

⟨⟩ Ingredients

1.5 kg frozen pork belly
300 g coarse salt

⟨⟩ Marinade

1 tbsp salt
1/2 tsp five-spice powder
1/2 tsp spice ginger powder
1 tbsp rose wine

⟨⟩ Tools

1 sheet aluminum foil
1 piece cotton string

燒
腩
仔

⟨⟩ Method

1. Defrost the pork belly in the lower chamber of the refrigerator. Cook in boiling water for about 10 minutes. Rinse with cold water. Wipe dry.

2. Pierce the skin of the pork belly with a set of needle or fork to let the entire skin filled with holes.

3. Turn over the pork belly. Make two cuts of about 1.5 cm deep. Spread with the rose wine. Then rub with the five-spice powder and spice ginger powder. Marinate for about 3 to 4 hours.

4. Fold the aluminum foil into a long strip. Surround the pork belly. The top of the aluminum foil should be about 2 cm higher than the pork belly. Tie with the cotton string.

5. Lay the coarse salt evenly on the surface of the pork belly, about 1 cm thick.

6. Put the whole pork belly on a baking tray. Bake in a preheated oven at 200°C for about 45 minutes. Remove.

7. Remove the coarse salt from the surface. Bake in the oven again on upper heat for about 15 minutes until the skin is golden and crispy. Allow it to cool down. Chop into pieces. Serve.

 零失敗技巧
Successful cooking skills

為何在肉面剻入兩刀？

將玫瑰露酒、五香粉及沙薑粉抹入切口內，讓調味徹底滲入五花腩肉，肉香撲鼻！

Why score the meat by making two cuts on the surface?

This is to allow rose wine, five-spice powder and spice ginger powder to penetrate the meat through the cuts, making it fragrant!

鋪上大量的粗鹽，肉會太鹹嗎？

絕對不會！因為用錫紙圍着腩肉周圍，粗鹽只會對皮層做成效果，絕不會令肉質有所影響。

Will the meat be too salty by covering with a great amount of coarse salt?

Absolutely not! The pork belly is surrounded by aluminum foil. Only the skin will be affected but not the meat.

粗鹽有何作用？

大量的粗鹽可吸取腩肉皮層的水分，烤焗後外皮更香脆可口！

What is the use of coarse salt?

Abundant salt can help absorb the moisture in the skin layer of the pork belly. The roasted skin will be crunchier.

為何用叉子戳於五花腩皮層？

五花腩用叉子戳後及烤焗，皮層卜脆酥化！

Why pierce the skin of the pork belly with a fork?

It is to make the roasted skin crispy and tasty!

芝心蝦丸

Cheese Stuffed Shrimp Balls

 容易指數：★★☆☆☆

材料

蝦仁 600 克
車打芝士 60 克（切成小粒）
生粉 1 湯匙

調味料

鹽 1/3 茶匙
生粉 1 茶匙
麻油及胡椒粉各少許

做法

1. 蝦仁用鹽 1/4 茶匙及生粉 2 茶匙擦勻，待半小時後，沖洗，瀝乾水分，用乾布吸乾。

2. 蝦仁用刀拍散，再用刀背輕剁。下調味料順一方向攪至起膠，撻入碗內數次，冷藏片刻。

3. 用水弄濕雙手，舀 1.5 湯匙蝦膠放於手掌，釀入芝士粒，搓成圓球狀，撲上少許生粉。

4. 放入熱油炸至金黃色，蘸沙律醬及茄汁享用。

1

2

3

4

◍ Ingredients

600 g shelled shrimp
60 g Cheddar cheese (finely diced)
1 tbsp caltrop starch

◍ Seasoning

1/3 tsp salt
1 tsp caltrop starch
sesame oil
ground white pepper

◍ Method

1. Rub the shelled shrimp with 1/4 tsp of salt and 2 tsp of caltrop starch. Leave for 1/2 hour. Rinse and drain. Use a dry cloth to wipe dry.

2. Pound the shelled shrimp with a knife. Chop gently with the back of the knife. Add the seasoning. Stir in one direction until sticky. Throw into a bowl for several times. Refrigerate for a moment.

3. Damp both hands. Put 1.5 tbsp of the minced shrimp in the palm. Stuff with the cheese. Knead into a ball. Coat with a little caltrop starch.

4. Deep-fry in hot oil until golden. Dip in salad sauce and ketchup to eat.

芝心蝦丸

◍ 零失敗技巧 ◍
Successful cooking skills

為何用刀背輕剁蝦仁？
不會折斷蝦的纖維，也容易拌成蝦膠！
Why chop the shelled shrimp gently with the back of a knife?
It will not break the meat fiber and will easily make into minced shrimp!

如何吃到爽滑彈牙的蝦丸？
將蝦膠多撻幾次，增加黏力與彈性，你會嘗到爽滑彈牙的蝦丸。
How do you make the shrimp balls springy?
Lift the minced shrimp and slap it back into the bowl forcefully for a few times. That helps make the minced shrimp sticky and the shrimp balls bouncy.

海南雞飯

Hainanese Chicken Rice

容易指數：★★★★☆

◎ 材料
嫩光雞 1 隻（約 2 斤至 2.5 斤）
紹酒 1 湯匙
白米 12 兩
蒜茸 3 湯匙
冰水適量

◎ 浸雞料
清水約 14 杯（3.5 公升）
粗鹽 1 1/4 湯匙
薑 2 片
葱 3 條

◎◎ 辣椒汁（拌勻）

指天椒碎半湯匙
蒜茸 1 茶匙
鹽 1/3 茶匙
雞粉 1/4 茶匙
熟油 1.5 湯匙

◎◎ 薑茸汁（拌勻）

幼薑茸 4 茶匙
鹽 1/3 茶匙
雞粉 1/4 茶匙
熟油 1.5 湯匙

◎◎ 做法

1. 光雞洗淨，抹乾水分，用紹酒抹勻雞腔待約 10 分鐘。
2. 燒滾浸雞料的清水，放入粗鹽、薑及葱，放入雞後又取出（重覆此步驟 3 次），再放入雞以慢火煮約 25 分鐘，關火，加蓋焗至雞全熟，取出，浸冰水，斬件，排於碟上，浸雞湯留用。
3. 白米洗淨，瀝乾水分，備用。
4. 熱鑊下油，下蒜茸用慢火炒香，加入白米炒至水分收乾，再轉放電飯煲內，注入適量浸雞湯及鹽半茶匙拌勻，煮熟。
5. 雞件蘸辣椒汁及薑茸汁，伴油飯品嘗。

◎◎ 零失敗技巧 ◎◎
Successful cooking skills

雞隻為何多次取出及浸入雞湯料？
令雞肉及雞腔慢慢暖透，再用慢火浸熟，以免雞肉一下子由凍受熱，破壞肉質。
Why repeat soaking in and removing the chicken from the sauce?
This is to let the meat and the inside of the chicken warm up slowly. Then soak and cook the chicken over low heat until done. It can prevent the meat texture from damaging by a sudden change of temperature from low to high.

可加強火力浸煮雞隻嗎？
用慢火長時間浸煮，雞肉嫩滑入口，別加強火力！
Can we use high heat to soak and cook the chicken?
Do not adjust to high heat! The chicken will be soft and gentle by soaking and cooking over low heat for a long time.

雞必須浸冰水嗎？
浸冰水後，雞皮爽滑無比！
Is it necessary to soak the chicken in iced water?
By doing so, the skin of the chicken will be extremely smooth and crunchy.

1

2

3

Ingredients

1 young chicken (about 1.2 kg to 1.5 kg)
1 tbsp Shaoxing wine
450 g rice
3 tbsp finely chopped garlic
iced water

Ingredients for soaking chicken

14 cups water (3.5 litres)
1 1/4 tbsp coarse salt
2 slices ginger
3 sprigs spring onion

Chilli sauce (mixed well)

1/2 tbsp chopped bird's eye chillies
1 tsp finely chopped garlic
1/3 tsp salt
1/4 tsp chicken bouillon powder
1.5 tbsp cooked oil

Grated garlic sauce (mixed well)

4 tsp finely grated garlic
1/3 tsp salt
1/4 tsp chicken bouillon powder
1.5 tbsp cooked oil

Method

1. Rinse the chicken. Wipe dry. Spread the Shaoxing wine evenly on the chicken cavity. Rest for about 10 minutes.

2. Bring the water used for soaking the chicken to the boil. Add the coarse salt, ginger and spring onion. Put in the chicken and then remove (repeat for 3 times). Put in the chicken again and cook over low heat for about 25 minutes. Turn off heat. Cover with a lid and rest until the chicken is fully done. Soak in iced water. Chop into pieces and arrange on a plate. Reserve the soaking sauce for later use.

3. Rinse the rice. Drain and set aside.

4. Add oil in a heated wok. Stir-fry the garlic over low heat until fragrant. Put in the rice and stir-fry until dry. Transfer to a rice cooker. Pour in some sauce from step 2 and 1/2 tsp of salt. Mix well. Cook the rice.

5. Serve the chicken with the rice, chilli sauce and grated garlic sauce.

魚茸餃兩吃

Fish Dumplings in Two Flavours

容易指數：★★☆☆☆

◎ 材料

扁鮫魚 1 條（約 1 斤 4 兩）
上海雲吞皮 24 塊
葱粒 2 湯匙
芫茜碎 4 湯匙

◎ 調味料

鹽 1.5 茶匙
胡椒粉少許
粟粉 2 湯匙
水 5 湯匙

◎ 做法

1. 扁鮫魚劏好，用刀切出兩片魚肉（魚頭及魚骨留用），洗淨。

2. 用廚房紙抹乾血水及水分，用大匙順魚肉逆紋刮出魚肉，用手挑淨小骨。（參考 p.9）

3. 將魚肉剁幼成魚茸，加入調味料順一方向拌勻，下芫茜碎拌成魚茸餡，撻入碗內數次至有膠質，在雲吞皮放入適量餡料包成魚茸雲吞。

4. 魚頭及魚骨洗淨，抹乾水分，放入油鑊內加薑片略煎，注入滾水 4 杯，用中火煲半小時，加少許鹽，隔去魚骨成魚湯，備用。

5. 燒滾清水半鑊，放入全部雲吞煮 8 分鐘至全熟，盛起，瀝乾水分。

6. 將半份雲吞放入魚湯煮成魚湯魚茸餃；另一半雲吞放入平底鑊煎香一邊，下葱粒至香，上碟享用。

Ingredients

1 Korean mackerel (about 750 g)
24 Shanghai style wonton wrappers
2 tbsp diced spring onion
4 tbsp chopped coriander

Seasoning

1.5 tsp salt
ground white pepper
2 tbsp cornflour
5 tbsp water

Method

1. Scale and gill the fish. Cut out fish fillet from two sides (reserve the head and bone). Rinse.

2. Wipe the meat dry. Follow the cross grain to scrape the meat off with a spoon. Pick off all tiny bones. (refer to p.9)

3. Chop the meat into puree. Stir in the seasoning in one direction. Mix well. Add the coriander and mix well as fish filling. Throw into a bowl for a few times until it is sticky. Wrap a small amount of the fish filling in a wonton wrapper as fish dumpling.

4. Rinse the fish head and bone. Wipe them dry. Fry them with ginger in a wok with oil. Pour in 4 cups of boiling water. Cook over medium heat for 1/2 hour. Add a little salt. Filter the fish soup.

5. Bring 1/2 wok of water to the boil. Put in all the dumplings. Cook for 8 minutes until fully done. Drain.

6. Put 1/2 portion of the dumplings into the fish soup. Cook for a while. Serve the fish dumplings in soup. Put the rest dumplings into a pan. Fry one side until fragrant. Add the spring onion and stir-fry until scented. Serve.

魚
茸
餃
兩
吃

零失敗技巧
Successful cooking skills

搭配不同魚打成魚茸，口感大有不同？

魚茸餃可依個人口味選配不同魚種：扁鮫魚＋狗棍魚——肉質幼滑、味鮮；扁鮫魚＋雜魚——肉質彈牙、味濃。

Does the mouthfeel differ if I use different fish for the filling?

You may choose different fish according to your preference. Use Korean mackerel and blacktail lizardfish for fine texture and umami. Use Korean mackerel and other assorted small fish for bouncy texture and rich flavours.

難以將魚肉攪打起膠，為甚麼？

刮魚肉前，必須徹底吸乾水分，水分太多難以拌至起膠，而且肉質偏腍。

Why is it hard to stir the minced fish into sticky paste?

Wipe the water thoroughly before scraping the meat off the fish. Too much moisture makes it hard to stir into sticky paste and also makes the meat too tender.

魚肉需要撻入碗內多少次？

要視乎魚茸的黏稠程度，見魚茸不黏碗底即可；若拌撻太久，魚茸欠軟滑的口感。若做魚蛋的話，可多撻幾次令肉質更具彈性。

How many times should the minced fish be thrown into a bowl?

It depends on the stickiness of the minced fish. It is done when it does not stick to the bowl. Long stirring and constantly throwing will make it rough to eat. You may throw it more if it is used to make spongy fish balls.

選用哪款雲吞皮？

我建議選用上海雲吞皮，口感滑嫩，是包裹魚茸餃最佳之選。

Use which type of wonton wrapper?

I suggest using Shanghainese wonton wrapper which tastes soft and smooth. It is perfect for fish dumplings.

包魚茸餃有何竅門？

別包入太多餡料，太滿容易弄破餃子皮，令餡料溢出，賣相不美！

What is the tip of wrapping the woton?

Do not stuff with too much filling! Otherwise, the wrapper will break and the filling will spill, making the dumplings look unappealing!

叉燒醬豬腩肉雙拼

Grilled Pork Cheek and Belly
in Char Siu Sauce

容易指數：★★★★☆

材料

豬腩肉 1 條（約 10 兩）
豬頸肉 1 塊
叉燒醬 4 湯匙（參考 p.16）
紹酒 2 湯匙
蜜糖適量

做法

1. 豬腩肉、豬頸肉洗淨，吸乾水分，加入紹酒及叉燒醬拌勻，醃 2 小時。

2. 焗盤墊上焗爐紙，排上豬腩肉及豬頸肉。

3. 預熱焗爐 10 分鐘，放入豬腩肉及豬頸肉，用 200℃焗 20 分鐘，翻轉豬腩肉及豬頸肉再焗 15 分鐘，掃上蜜糖再焗 5 分鐘，取出切塊享用。

Ingredients

1 piece pork belly (about 375 g)
1 piece pork cheek
4 tbsp homemade Char Siu Sauce (refer to p.16)
2 tbsp Shaoxing wine
honey

Method

1. Rinse the pork belly and cheek. Wipe dry. Add Char Siu Sauce and Shaoxing wine. Mix well. Leave them for 2 hours.

2. Line a baking tray with baking paper. Put in the pork belly and cheek.

3. Preheat an oven to 200℃ for 10 minutes. Bake the pork belly and cheek for 20 minutes. Flip them upside down to grill the other side for 15 more minutes. Brush on honey and bake for 5 more minutes. Cut into pieces and serve.

1

2

3

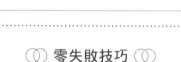

⦾ 零失敗技巧 ⦾
Successful cooking skills

為何加紹酒醃味？
令肉類加添酒香味，去除肉類的異味。
Why do you marinate the pork with Shaoxing wine?
It covers up the gamey taste of the pork and adds a lovely fragrance.

焗爐用上火還是下火烤焗？
烤焗時開啟上下火，令豬腩肉兩面受熱均勻，香氣四溢。
Do you turn on both the top and bottom heat filament in the grilling process?
Yes, you should turn on both top and bottom heat in the oven, so that the pork gets cooked evenly from all directions.

烤焗後的豬腩肉肥膩嗎？
絕不肥膩，經高溫烤焗的肥肉，分泌了大量油脂，反而肉質爽口彈牙。
Will the pork belly be greasy?
No, it won't. The grease is cooked and drained out of the meat in the high-temperature grilling process. The meat will be juicy and has a lovely chew to it.

雞茸豆腐球

Deep Fried Chicken and Tofu Balls

容易指數：★★★☆☆

◎ 材料

布包豆腐 1 塊
圓形豆腐泡 12 個
雞肉 3 兩
蔥粒 1 湯匙

◎ 醃料

生抽半湯匙
糖半茶匙
胡椒粉少許
粟粉 1 茶匙

◎ 蘸汁

大紅浙醋 1 小碟

◎ 做法

1. 雞肉洗淨，剁成茸，下醃料拌勻，炒熟備用。

2. 布包豆腐壓成茸，加入雞肉及蔥粒拌成餡料。

3. 豆腐泡剪開一小角，用手輕輕反向後（白色向外），將適量雞茸豆腐餡釀入中空位置。

4. 燒滾油，下雞茸豆腐球用中火炸至金黃色，瀝乾油分，伴蘸汁享用。

◎ Ingredients

1 cloth-wrapped tofu
12 round deep-fried tofu puffs
113 g chicken meat
1 tbsp chopped spring onions

◎ Marinade

1/2 tbsp light soy sauce
1/2 tsp sugar
ground white pepper
1 tsp cornflour

◎ Dipping sauce

1 small plate red vinegar

◎ Method

1. Rinse the chicken meat. Chop into puree. Mix well with the marinade. Stir-fry until done. Set aside.

2. Mash the tofu. Add the chicken meat and spring onions. Mix well as filling.

3. Cut a small corner of the tofu puffs. Slowly flip the inside over (with the white side facing outwards). Stuff some of the fillings in the middle to become a ball.

4. Heat oil in a wok. Deep-fry the balls over medium heat until golden brown. Drain well. Serve with the dipping sauce.

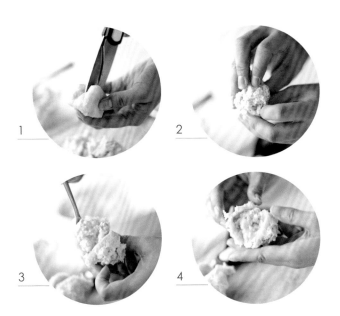

1

2

3

4

◎◎ 零失敗技巧 ◎◎

Successful cooking skills

可用牛肉代替雞肉嗎？

可用免治牛肉代替，省卻剁肉的步驟。

Can we use beef instead?

To save the time of chopping step, you can use minced beef.

釀入餡料後立即下油鍋炸嗎？

建議放一會才炸，避免餡料容易溢出。

Is it to deep-fry the ball right after it is stuffed?

Set aside for a while before deep-frying to avoid the filling coming out.

豆腐球需要炸兩次嗎？

不需要，因豆腐泡已炸了一遍，減去部份水分。

Is it necessary to deep-fry the ball twice?

No, the tofu puff has been deep-fried reducing part of the water.

甘蔗雞

Chicken Sugar Cane Skewers

容易指數：★★☆☆☆

◎ **材料**
雞胸肉 8 兩
去皮馬蹄 4 粒
竹蔗 2 段（每段約 12 厘米長）
生粉適量

◎ **調味料**
鹽 1/4 茶匙
生抽 2 茶匙
糖 1/4 茶匙
胡椒粉少許
麵包糠 3 湯匙

◎ **蘸汁（拌勻）**
青檸汁 1.5 湯匙
魚露 2 湯匙
指天椒碎 1 茶匙
蒜茸半湯匙
白醋 1 茶匙
糖 2.5 湯匙
鹽 1/4 茶匙
凍開水 1 湯匙

◎ **做法**

1. 馬蹄切幼粒，吸乾水分；竹蔗去皮，每段直破開為 4 份，共 8 支。

2. 雞胸肉剁成茸，加入馬蹄粒及調味料，順一方向拌至起膠，分成 8 份餡料。

3. 取一份餡料，壓平，放上竹蔗段，捏實，頭尾兩端露出竹蔗段。

4. 均勻地沾上生粉，搓圓，放入熱油炸熟，蘸汁料享用。

◎ **零失敗技巧** ◎
Successful cooking skills

包裹雞肉茸時，有何技巧？
先用凍開水弄濕雙手，令雞茸不容易黏手，包捏容易，快速完成。
How to wrap the sugar cane in the minced chicken skilfully?
Wet your hands with cold drinking water, the minced chicken do not stick on hands when wrapping. It will then be done easily and quickly.

調味料必須拌入麵包糠嗎？
麵包糠增加雞茸料的黏合度，令雞肉不容易散開來。
Is it necessary to mix bread crumbs with the seasoning?
The bread crumbs will enhance the stickiness of the minced chicken, making it hardly to break up.

1

2

3

4

Ingredients

300 g chicken breast
4 peeled water chestnuts
2 sections sugar cane (about 12 cm long for each section)
caltrop starch

Seasoning

1/4 tsp salt
2 tsp light soy sauce
1/4 tsp sugar
ground white pepper
3 tbsp bread crumbs

Dipping sauce (mixed well)

1.5 tbsp lime juice
2 tbsp fish sauce
1 tsp chopped bird's eye chillies
1/2 tbsp finely chopped garlic
1 tsp white vinegar
2.5 tbsp sugar
1/4 tsp salt
1 tbsp cold drinking water

Method

1. Finely dice the water chestnuts. Wipe dry. Peel the sugar cane. Quarter each section lengthwise to make a total of 8 skewers.

2. Chop the chicken breast into puree. Add the water chestnuts and seasoning. Stir in one direction until sticky. Divide into 8 portions.

3. Take a portion of the chicken mixture. Flatten. Put a sugar cane skewer on top. Wrap the skewer in the chicken mixture tightly with both ends untouched.

4. Evenly coat with the caltrop starch. Knead into a round shape. Deep-fry in hot oil. Serve with the dipping sauce.

露筍雞髀菇炒鴿片

Stir-fried Pigeon with Asparagus and King Oyster Mushroom

容易指數：★☆☆☆☆

⬤⬤ 材料

冰鮮乳鴿 1 隻
雞髀菇 1 個（切片）
露筍 3 條
紅蘿蔔半個（切片）
薑 2 片
乾葱茸 2 茶匙
XO 醬 1 湯匙

⬤⬤ 醃料

鹽及糖各半茶匙
生抽、紹酒及粟粉各 1 茶匙
麻油及胡椒粉各少許

⬤⬤ 調味料

鹽及糖半茶匙
生抽 1 茶匙
麻油及胡椒粉各少許
水 3 湯匙
粟粉 1 茶匙

⬤⬤ 做法

1. 乳鴿洗淨，抹乾水分，開邊，沿骨將鴿肉起出，切片，用醃料拌勻。

2. 露筍用小刀削去外皮，飛水，斜切成段。

3. 燒熱油 1 湯匙，加入薑片及乾葱茸炒香，下鴿片略煎兩面，再推散略炒，灑入 XO 醬炒勻。

4. 加入露筍、雞髀菇及紅蘿蔔炒片刻，最後下調味料拌勻即成。

露筍雞髀菇炒鴿片

⬤⬤ Ingredients

1 chilled pigeon
1 king oyster mushroom (sliced)
3 stalks asparagus
1/2 carrot (sliced)
2 slices ginger
2 tsp finely chopped shallot
1 tbsp XO sauce

⬤⬤ Marinade

1/2 tsp each of salt and sugar
1 tsp each of light soy sauce, Shaoxing wine and cornflour
sesame oil
ground white pepper

⬤⬤ Seasoning

1/2 tsp each of salt and sugar
1 tsp light soy sauce
sesame oil
ground white pepper
3 tbsp water
1 tsp cornflour

⬤⬤ Method

1. Rinse the pigeon. Wipe dry. Cut along the edge. Remove the meat along the bones. Cut into slices. Mix with the marinade.

2. Peel the asparagus with a knife. Scald and cut diagonally into sections.

3. Heat up 1 tbsp of oil. Stir-fry the ginger and shallot until fragrant. Add the pigeon and fry both sides slightly. Scatter and stir-fry slightly. Sprinkle with the XO sauce and stir-fry evenly.

4. Put in the asparagus, king oyster mushroom and carrot. Stir-fry for a moment. Add the seasoning and mix well. Serve.

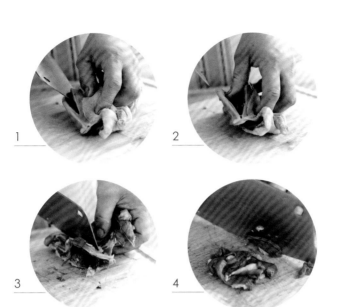

◎ 零失敗技巧 ◎
Successful cooking skills

乳鴿很難起肉嗎？
由於乳鴿體型小，起肉確實有點難度，建議用一把鋒利的刀試試，必定靈活做到。

It is difficult to take the meat off the pigeon, isn't it?

Yes, it is. As the pigeon is quite small in size, I suggest trying with a sharp knife. It must be easier.

如何吃到爽嫩的露筍？
除了削去外皮，亦可於露筍近末端處折斷，去掉粗硬的末段即可。

How to make asparagus soft and crunchy?

Besides peeling the asparagus, you can break its end to remove the tough part.

市場沒有雞髀菇，可用甚麼代替？
秀珍菇、鮑魚菇或蘑菇等皆可，但雞髀菇肉質厚，有嚼勁！

King oyster mushroom cannot be found in the market. What is the substitute?

Oyster, abalone or button mushrooms can be used, but king oyster mushroom is meaty and chewy.

咕嚕百花釀油條

Shallow-fried Stuffed Dough Stick
with Minced Prawn

容易指數：★★☆☆☆

◯◯ **材料**
蝦肉 4 兩
油條 1 孖
西芹粒 2 湯匙

◯◯ **醃料**
鹽 1/8 茶匙
胡椒粉少許

◯◯ **蘸汁**
水 4 湯匙
白醋 2 茶匙
糖 1 湯匙
茄汁 1 湯匙
生粉半茶匙

◯◯ **做法**

1. 油條撕成兩條，剪成約 2 厘米短度，備用。

2. 蝦肉去腸，洗淨，抹乾水分，用刀拍打成蝦膠，加入醃料及西芹粒順一方向攪至帶黏性。

3. 掏出油條的軟麵糰部分，釀入適量蝦膠。

4. 燒熱適量油，下百花油條半煎炸至餡熟皮脆，盛起。

5. 煮滾蘸汁，伴百花油條享用。

◯◯ Ingredients

150 g shelled prawns
1 pair deep-fried dough stick
2 tbsp diced celery

◯◯ Marinade

1/8 tsp salt
ground white pepper

◯◯ Dipping sauce

4 tbsp water
2 tsp white vinegar
1 tbsp sugar
1 tbsp ketchup
1/2 tsp caltrop starch

◯◯ Method

1. Tear the dough stick into two pieces and cut into sections of about 2 cm long. Set aside.

2. Devein shelled prawns. Rinse and wipe dry. Pat with the flat side of a knife to become the minced prawn. Add marinade and diced celery. Stir in one direction until sticky.

3. Scrape out the soft part from dough stick and stuff in the minced prawn.

4. Heat oil in a wok. Shallow-fry the minced prawn stuffed dough stick until the fillings are done and the skin is crispy. Drain.

5. Bring the dipping sauce to the boil. Serve with the stuffed dough stick.

◯◯ 零失敗技巧 ◯◯
Successful cooking skills

百花釀油條可預先釀妥嗎？
絕對可以，油條釀妥後冷藏，享用時炸脆，簡單方便。
Can I stuff the minced prawn in advance?
Yes, you can stuff the minced prawn into the dough stick in advance and store in the refrigerator. Only fry it when going to serve. It is simple and convenient.

打蝦膠有何注意之處？
剁碎及拌蝦膠用之器皿必須徹底乾淨；拌蝦膠時順一方向用力攪拌，一定成功打出蝦膠。
What should be noted about when stirring the minced prawn?
The tools for chopping shelled prawns and the container for mixing minced prawn should be clean. Also stir in one direction in force can give minced prawn successfully.

如何確知蝦肉餡料熟透？
見蝦肉轉成紅色及肉質結實，即代表蝦肉熟透。
How to know if the prawn fillings are done?
When the prawn flesh turns red and becomes firm in texture, it is done.

田園芝士豆腐批

Vegetables and Tofu Pie with Cheese

容易指數：★★★☆☆

材料

布包豆腐 1 塊
西蘭花 1/4 個
車厘茄 6 粒
煙肉 2 條（切粒）
洋葱碎 3 湯匙
雞蛋（大）3 個
巴馬臣芝士絲 2 湯匙

調味料

鹽半茶匙
糖 1/3 茶匙
胡椒粉少許

做法

1. 西蘭花切成小朵，洗淨，飛水備用。

2. 車厘茄去蒂，洗淨，切粒。

3. 布包豆腐用叉子壓成茸；雞蛋拌成蛋液。

4. 洋葱及煙肉炒香，盛起，與其他材料（巴馬臣芝士粉除外）及調味料拌勻。

5. 燒熱平底鑊，下油 2 湯匙，傾入步驟（4）的漿料，煎成兩面金黃色及凝固的厚批。

6. 灑上巴馬臣芝士絲，連鑊放入已預熱之焗爐，用 180℃ 焗至芝士絲開始溶化，取出，待片刻，切塊享用。

Ingredients

1 cloth-wrapped tofu
1/4 broccoli
6 cherry tomatoes
2 rashers bacon (diced)
3 tbsp chopped onion
3 eggs (large)
2 tbsp shredded Parmesan cheese

Seasoning

1/2 tsp salt
1/3 tsp sugar
ground white pepper

Method

1. Cut the broccoli into small florets. Rinse and scald. Set aside.

2. Remove the stalks of the cherry tomatoes. Rinse and dice.

3. Mash the tofu with the fork. Whisk the eggs.

4. Stir-fry the onions and bacon until fragrant. Dish up. Mix well with the other ingredients (except the Parmesan cheese) and the seasoning.

5. Heat a pan. Put in 2 tbsp of oil. Pour in the mixture from step 4. Fry until both sides turn golden brown to become a firm deep-dish pie.

6. Sprinkle the Parmesan cheese on top. Put the pan into a preheated oven. Bake at 180°C until the cheese starts to melt. Remove and set aside for a while. Cut into pieces and serve.

零失敗技巧
Successful cooking skills

為何選用布包豆腐？
布包豆腐質地幼滑，而且價錢便宜。
Why choose cloth-wrapped tofu?
It is silky and economical.

蔬菜容易溢出水分，豆腐批會滲有水分嗎？
將蔬菜料慢慢烘乾及煮熟，以免過多水分溢出，影響口感。
Will water released from vegetables permeate the pie?
To avoid excessive water affecting the taste, slowly dry the vegetables on heat until fully done.

菠菜番茄火腿餡餅

Spinach and Tomato Quiche

容易指數：★★☆☆☆

＊可製成 5 個（10 厘米 x 7.5 厘米）餡餅

 批皮材料

凍牛油 75 克
麵粉 150 克
蛋黃 1 個
冰水約 4 湯匙
鹽半茶匙

 餡料

菠菜 150 克
洋葱碎半杯
火腿 80 克
羊奶芝士 100 克
番茄 1 個
橄欖油 2 茶匙
雞蛋 2 個
忌廉 1 杯
鹽半茶匙，黑椒粉少許

準備工夫

1. 批皮：牛油切小粒；麵粉、鹽篩勻。牛油粒加入麵粉內，用手指擦成麵包糠狀，加入蛋黃和適量冰水搓成粉糰。
2. 菠菜洗淨，用熱水拖軟，過冷河，切碎，擠乾水分。
3. 洋葱切碎，用橄欖油炒軟。
4. 番茄切片；火腿切碎；芝士刨碎。
5. 蛋和忌廉拌勻，加入鹽和黑椒粉拌勻備用。
6. 焗爐調至 190℃，預熱 10 分鐘。

做法

1. 將粉糰擀薄，分別放於批盤內，用叉刺孔，放入雪櫃雪片刻。
2. 放入已預熱的焗爐內焗約 15 分鐘，取出。
3. 將菠菜碎、番茄、洋葱、火腿拌勻，倒入忌廉蛋液，拌勻。
4. 將混合料舀進批皮內，灑上芝士碎放入已預熱焗爐焗約 25 分鐘即成。

零失敗技巧
Successful cooking skills

批皮酥脆的竅門是甚麼？
牛油粒要凍，水亦要冰凍，同時不要過量搓揉麵糰。

Is there any trick to make the crust crispier?
The butter should cold enough and the water should be iced. Do not over work the dough.

為甚麼批皮要放在雪櫃雪一會才烘焙？
讓批皮鬆弛，以免烘焙時收縮。同時批皮的牛油不會溶掉，而影響鬆脆度。

Why did you refrigerate the pie crust for a while before baking?
That would allow time for the crust to rest and it won't shrink as much when baked. Besides, the butter in the crust will not melt this way, so that the crust will be crispier.

*makes five 10 cm x 7.5 cm quiches

◎ Pie crust

75 g butter (chilled)
150 g flour
1 egg yolk
4 tbsp iced water
1/2 tsp salt

◎ Filling

150 g spinach
1/2 cup chopped onion
80 g cooked ham
100 g Feta cheese
1 tomato
2 tsp olive oil
2 eggs
1 cup whipping cream
1/2 tsp salt
ground black pepper

菠
菜
番
茄
火
腿
餡
餅

◎ Preparation

1. To make the crust, dice the butter finely. Sift flour and salt together. Put diced butter into the dry ingredients. Rub flour and salt into the butter with your fingertips until it resemble breadcrumbs. Add egg yolk and iced water. Knead into dough.

2. Rinse the spinach and blanch in boiling water. Rinse with cold water. Chop it up and squeeze dry.

3. Chop the onion. Stir fry in olive oil until transparent.

4. Slice the tomato. Chop the ham. Grate the cheese.

5. Mix eggs and whipping cream together. Add salt and ground black pepper. Stir well.

6. Preheat an oven to 190°C for 10 minutes.

◎ Method

1. Roll out the dough and press into individual pie pans. Prick holes on them with a fork. Refrigerate for a while.

2. Blind bake the crust in the preheated oven for 15 minutes. Leave them to cool.

3. To make the filling, mix together spinach, tomato, onion and ham. Pour in the cream and egg mixture. Stir well.

4. Pour the filling mixture into the baked pie crusts. Sprinkle with the grated cheese. Bake in the preheated oven for 25 minutes. Serve.

芝士蝴蝶酥

Cheese Palmier

 容易指數：★★☆☆☆

＊可製成 40 件

◎ 材料

急凍酥皮 1 包（解凍）
麵粉 2 湯匙
雞蛋 1 個
合桃碎適量
Brie 芝士 100 克

◎ 做法

1. 芝士切成幼粒；合桃切碎。

2. 灑少許麵粉於工作枱，將酥皮擀薄成厚約 4 毫米、25 厘米 x25 厘米的正方形。

3. 酥皮抹上蛋液，將芝士、合桃碎平均地鋪於酥皮上，從上下兩邊向內捲起，相疊。

4. 放入冰格雪約 1 小時，取出，切約半厘米厚。

5. 放於鋪上牛油紙的焗盤內，放入已預熱焗爐用 200℃焗約 18 分鐘，取出，攤涼後享用。

芝士蝴蝶酥

＊ makes 40 pastries

◎ Ingredients

1 pack frozen puff pastry
2 tbsp flour
1 egg
chopped walnuts
100 g Brie cheese

◎ Method

1. Dice the cheese finely. Chop the walnuts.

2. Sprinkle some flour on the counter. Roll out the puff pastry into a 25 cm square, about 4mm thick.

3. Brush egg wash on the puff pastry. Arrange the diced cheese and chopped walnuts evenly over the pastry. Roll from both ends toward the centre.

4. Refrigerate for 1 hour. Slice into half cm thick pieces.

5. Arrange on a baking tray or cookie sheet lined with baking paper. Bake in the preheated oven at 200℃ for 18 minutes. Leave it to cool and serve.

1

2

3

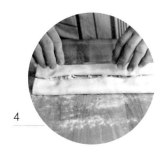

4

5

6

◎ 零失敗技巧 ◎
Successful cooking skills

有甚麼成敗關鍵的竅門？
捲酥皮時不要捲得太實，宜鬆手，這可提供足夠空間讓酥皮膨脹。
Is there any trick to this recipe?
Just don't roll the puff pastry too tightly. Otherwise, the pastry won't have room for puffing up and you'd end up with dense palmiers, not fluffy ones.

用 200℃的爐溫，不會太高嗎？
用比較高的爐溫，蝴蝶酥才容易鬆起。
Baking the cheese palmier at 200℃ , is it a bit too hot?
No, it's not. Actually, the puff pastry tends to rise and puff up better at high temperature.

朱古力花生醬扭紋蛋糕

Peanut Butter Chocolate Marble Muffins

容易指數：★★★☆☆

＊可製成 12 個份量

◎ 材料
黑朱古力 75 克
麵粉 150 克
發粉 1.5 茶匙
粗粒花生 150 克
鹽 1/4 茶匙
牛油 150 克
幼砂糖 100 克
黃幼砂糖 50 克
雞蛋 3 個
鮮奶 75 毫升

◎ 餅面
牛油 1 湯匙
花生醬 1 湯匙
黑朱古力 90 克
蜜糖 1 湯匙

◎ 做法
1. 麵粉、發粉及鹽篩勻。
2. 黑朱古力切碎，座於熱水至溶，待涼。
3. 牛油、幼砂糖及黃砂糖打透至忌廉狀，加入雞蛋略打，下粗粒花生拌勻。
4. 將步驟（1）的粉料及鮮奶，分兩次拌入上述混合物。
5. 混合物分成兩份，其中一份與朱古力漿拌勻。
6. 鬆餅盆墊上烘焙紙，用湯匙分別將兩份漿料放入模內，用叉輕輕攪動成雲石花紋。
7. 放入預熱焗爐，以 190℃ 焗約 25 分鐘，取出待涼。
8. 餅面材料座於熱水內，澆於餅面即可。

◎ 零失敗技巧 ◎
Successful cooking skills

朱古力加熱時有何注意之處？
朱古力座於熱水時，緊記切勿滲有水分。
What is the techniques of melting the chocolate?
When you melt the chocolate over a pot of simmering water, make sure you don't accidentally get any water into the chocolate.

如何令雲石效果更美觀？
用叉或牙籤輕柔地在漿料轉動，雲石效果會慢慢地顯露出來。
How do you make the perfect marbling?
Gently swirl the mixture with a fork or a toothpick. The marbling will show gradually.

*makes 12 muffins

 Ingredients

75 g dark chocolate
150 g flour
1.5 tsp baking powder
150 g peanuts
1/4 tsp salt
150 g butter
100 g castor sugar
50 g brown castor sugar
3 eggs
75 ml milk

 Glazing

1 tbsp butter
1 tbsp peanut butter
90 g dark chocolate
1 tbsp honey

朱古力花生醬扭紋蛋糕

Method

1. Sieve flour, baking powder and salt together.

2. Finely chop the dark chocolate. Melt them in a bowl over a pot of simmering water.

3. Beat butter, castor sugar and brown castor sugar until pale. Add eggs and beat until well incorporated. Add the peanuts and mix well.

4. Add half of the dry ingredients from step 1 and half portion of milk Stir well. Add the remaining dry ingredients and milk and mix well.

5. Divide the batter into two equal portions. Add the melted dark chocolate into one portion and mix well.

6. Line the muffin tins with baking paper. Use a spoon to put some chocolate batter and some plain batter into the muffin tins. Stir with a fork gently to create marbling pattern.

7. Bake in a preheated oven at 190°C for about 25 minutes. Leave them to cool.

8. To make the glazing, mix all ingredients in a mixing bowl. Heat the bowl over a pot of simmering water until ingredients melt. Pour the glazing over the muffins. Serve.

豆渣果仁脆曲奇

Soybean Pulp and Nut Cookies

容易指數：★★★★☆

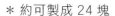

＊ 約可製成 24 塊

◎ 材料

已烘乾豆渣 100 克 ＊
低筋麵粉 150 克
黃砂糖 150 克
牛油 150 克（放於室溫回軟）
雞蛋 1 個
杏仁片 100 克
朱古力粒 50 克

◎ 豆渣做法

自製豆漿時，擠壓豆漿後，餘下的豆渣可冷藏 2 至 3 日，或用白鑊烘香儲存。

◎ 做法

1. 預熱焗爐至 170℃。

2. 牛油及砂糖打至忌廉狀，加入雞蛋打起，放入杏仁片拌勻。

3. 篩入低筋麵粉輕拌，再加入烘乾豆渣拌勻，最後下朱古力粒拌勻，分成 2 份麵糰，用保鮮紙包成球狀，冷藏約 3 小時至麵糰硬身。

4. 麵糰用擀麵棒壓成 5 毫米厚，用曲奇模切出所需形狀，放在鋪有牛油紙之焗盤，放入焗爐焗 18 至 20 分鐘至金黃色，取出，待涼即成。

豆渣果仁脆曲奇

*makes 24 cookies

◎ Ingredients

100 g bake-dried soybean pulp *
150 g cake flour
150 g brown castor sugar
150 g butter (soften at room temperature)
1 egg
100 g almonds flakes
50 g chocolate cubes

◎ Method for soybean pulp

Obtained by pressing soybean milk out. It can be preserved in a refrigerator for 2 to 3 days, or kept by stir-frying without oil.

◎ Method

1. Preheat an oven to 170℃.

2. Whisk the butter and sugar until creamy. Add the egg and stir until stiff. Put in the almonds and mix well.

3. Sieve the cake flour in the butter mixture and stir lightly. Add the soybean pulp and mix well. Mix in the chocolate cubes. Divide the dough into halves. Use a cling film to wrap and shape the dough into a ball. Refrigerate for about 3 hours until firm.

4. Roll out the dough until it is 5 mm thick with a rolling pin. Use cookie moulds to cut out the desired shapes. Put on a baking tray lay with baking paper. Bake for 18-20 minutes until golden brown. Set aside to cool.

◎◎ 零失敗技巧 ◎◎
Successful cooking skills

加入豆渣烘焗的曲奇，食味有何不同？
曲奇會更香更脆，以及散發陣陣濃濃的豆香味。
How is the flavour different by baking cookies with soybean pulp?
They are crunchier and smell fantastic with a strong soybean flavour.

如何做出鬆脆的豆渣曲奇？
豆渣必須徹底烘乾或炒乾，以免豆渣太濕潤，令曲奇不鬆脆。
How to make the cookies crunchy?
Fully dry the soybean pulp by baking or stir-frying; otherwise the moisture will make the cookies soft.

為何麵糰要冷藏至硬？
以免牛油溶化令麵糰失去彈性，影響塑形之效果。
Why refrigerate the dough until firm?
When butter melts, the dough loses its elasticity and is hard to shape.

迷你藍莓果仁月餅

*Mini Mooncakes with
Blueberry and Assorted Nuts*

容易指數：★★☆☆☆

◎ 材料

皮料

麵粉 150 克
糖膠 100 克
粟米油 1/4 杯
鹼水 1/4 茶匙
濃普洱茶 1 湯匙

◎ 餡料

合桃 60 克
杏仁 50 克
榛子 40 克
夏威夷果仁 40 克
糖冬瓜 50 克
蜜餞果皮 20 克
藍莓乾 50 克
粟米油 25 毫升
凍滾水 45 毫升
糖霜 50 克
糕粉 30 克
冧酒 1 湯匙

◎ 塗料

蛋黃 2 個
粟米油 1 茶匙
水 1 茶匙（與糖 1 茶匙調勻）
＊調勻所有材料

◎ 做法

餡料

1. 焗香果仁，切碎。

2. 所有材料（除糕粉）拌勻，最後加入糕粉拌勻；餡料搓成每個約 30 克圓粒。

◎ 綜合做法

1. 餅皮：麵粉放在深碗內，加入糖膠、油、濃普洱茶、鹼水拌勻，搓成粉糰，靜置 1 小時。

2. 將粉糰搓成長條形，再分割成每份約 15 克的小粉糰。

3. 將小粉糰壓薄，包入餡料成月餅的雛形，然後放入已灑粉的餅模中，按壓，脫去餅模。

4. 月餅上噴水，放入已預熱 210℃ 的焗爐內焗約 8 分鐘，取出月餅，餅面塗上蛋液，轉用 180℃ 焗約 10 分鐘，取出掃蛋液，再焗 10 分鐘即成。

Ingredients

Pastry skin

150 g flour
100 g pectin
1/4 cup corn oil
1/4 tsp food-grade lye
1 tbsp strong Pu-er tea

Filling

60 g walnuts
50 g almonds
40 g hazelnuts
40 g macadamia nuts
50 g candied wintermelon
20 g candied citrus peels
50 g dried blueberries
25ml corn oil
45ml cold drinking water
50 g icing sugar
30 g fried glutinous rice flour
1 tbsp rum

Egg wash (mixed well)

2 egg yolks
1 tsp corn oil
1 tsp water (mixed with 1 tsp of sugar first)

Method

Filling

1. Bake all nuts briefly. Finely chop them.

2. Mix all ingredients together (except fried glutinous rice flour). Stir in the fried glutinous rice flour at last. Stir well. Roll the filling with your hands into balls about 30 g each.

迷你藍莓果仁月餅

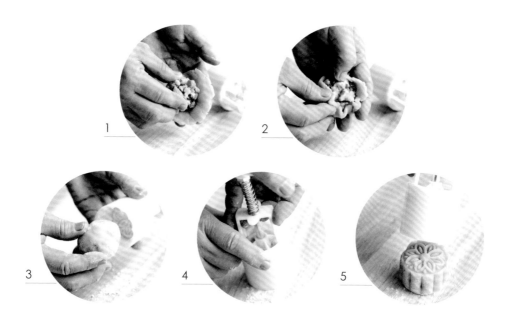

1

2

3

4

5

 Assembly

1. To make the pastry skin, put flour into a deep mixing bowl. Add pectin, oil, Pu-er tea and lye. Mix well and knead into dough. Leave it for 1 hour.

2. Roll the dough into long cylinder. Cut into short sections of dough about 15 g each.

3. Roll out each small piece of dough. Wrap in a ball of filling. Seal the seam well. Press the stuffed dough firmly into a cake mould. Tap the mould a few times to release the mooncake. Repeat this step until all ingredients are used up.

4. Spray drinking water on the mooncakes. Bake in a preheated oven at 210°C for about 8 minutes. Brush egg wash over the mooncakes. Turn the oven down to 180°C and bake for 10 more minutes. Brush a second layer of egg wash on them. Bake for 10 more minutes at last and serve.

零失敗技巧
Successful cooking skills

烘焗月餅時，有何需要注意？
放入焗爐前，緊記噴上水點，以免月餅太乾太硬。
Is there anything that needs my attention when I bake the mooncakes?
Sprinkle water on the mooncakes before baking. Otherwise, they may turn too dry and too hard.

如何挑選餡料？
任何果仁都可製成餡料，只要自己喜歡的即可，悉隨尊便。
How do you choose the filling ingredients?
You can use any nut as the filling. Just choose any type you like.

月餅形狀容易按壓出來嗎？
使用輕便的按壓式月餅模，簡單方便，輕輕一按，月餅花紋清晰明顯。
Is it easy to create the embossment on the mooncakes?
Use a hand-pressed mooncake mould for convenience. Just press it with your hand and the mooncakes will turn out perfect with beautiful embossment.

鮮雜果杏仁豆腐

Almond Tofu Pudding with Assorted Fruits

容易指數：★★★☆☆

◎ 杏仁豆腐材料

冷凍淡豆漿 1 杯 *
果凍粉 2 茶匙
砂糖 2 茶匙
杏仁油 1 滴

◎ 配料

鮮雜果適量
糖漿 1 湯匙

◎ 淡豆漿材料

黃豆 1 斤
水 20 杯（約 5 公升）

◎ 淡豆漿做法

1. 黃豆洗淨，用水浸泡 8 小時（浸過面約 3 吋），洗淨。

2. 在攪拌機內，放入一份黃豆及一份水（即黃豆半杯及水 1 杯），用高速磨成滑豆漿，以布袋過濾，徹底壓出豆漿，隔渣（豆渣可留用），重覆此步驟數次，直至全部完成。

3. 將豆漿及餘下的水分，傾入煲內滾起，轉慢火再煲 10 分鐘即可。

◎ 杏仁豆腐做法

1. 果凍粉及砂糖拌勻，拌入淡豆漿。

2. 燒滾豆漿混合料至微滾，期間不斷拌勻至果凍粉溶化，加入杏仁油略拌，傾入器皿內待涼，冷藏 2 小時至凝固成杏仁豆腐。

◎ 綜合做法

1. 鮮雜果切粒，備用。

2. 杏仁豆腐切粒，盛於碗內，加入鮮雜果，澆上糖漿享用。

淡豆漿做法 Method for soybean milk

◯◯ **Ingredients for almond tofu pudding**

1 cup light soybean milk (cold)*
2 tsp gelatin powder
2 tsp sugar
1 drop almond essence

◯◯ **Condiments**

fresh assorted fruits
1 tbsp syrup

◯◯ **Ingredients for soybean milk**

600 g soybeans
20 cups water (about 5 litres)

◯◯ **Method for soybean milk**

1. Rinse the soybeans. Soak in water for 8 hours (the height of water is about 3 inches above the surface of the soybeans). Rinse.

2. Put 1 part of soybeans and 1 part of water (1/2 cup of soybeans and 1 cup of water) in a blender. Blend at high speed until it becomes smooth soybean milk. Filter with a cloth bag. Fully press the soybean milk out. Strain off soybean pulp (reserve the soybean pulp). Repeat the steps until all the soybeans are finished blending.

3. Boil the soybean milk and the remaining water in a saucepan. Then turn to low heat and cook for 10 minutes.

鮮雜果杏仁豆腐

杏仁豆腐做法 Method for almond tofu pudding milk

○○ Method for almond tofu pudding

1. Mix the gelatin powder and sugar well. Mix in the light soybean milk.

2. Cook the soybean milk mixture until it boils slightly. Meantime, keep stirring until the gelatin powder dissovles. Add the almond essence and stir well. Pour into a container to cool. Refrigerate for 2 hours until firm.

○○ Assembly

1. Dice the assorted fruits. Set aside.

2. Dice the almond tofu pudding. Put in a bowl. Add the assorted fruits. Serve with the syrup.

○○ 零失敗技巧 ○○
Successful cooking skills

做杏仁豆腐必須用冷豆漿嗎？
用冷豆漿拌入果凍粉，不會出現顆粒狀。

Must we use cold soybean milk to make almond tofu pudding?

No grains will appear by mixing cold soybean milk with gelatin powder.

製作杏仁豆腐有何竅門？
果凍粉、砂糖及豆漿宜徹底拌勻，而且果凍粉必須煮至完全溶化，以免有顆粒塊。

What's the techniques of making the almond tofu pudding?

The gelatin powder, sugar and soybean milk must be mixed thoroughly. The gelatin powder must be cooked until it fully melts.

一滴杏仁油已足夠？
由於杏仁油香氣濃郁，故加一滴已香氣豐足。若喜歡濃濃的杏仁香味，不妨隨喜好酌加適量的杏仁油，但杏仁油過多，會帶苦澀味。

Is it enough to use only a drop of almond essence?

It is enough because almond essence has a strong smell. You can add more if you prefer it stronger, but too much essence will give a bitter taste.

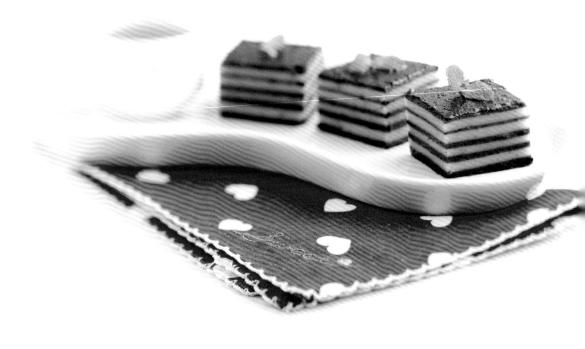

黑芝麻杏仁千層糕

Black Sesame and Almond Layer Cake

容易指數：★★☆☆☆

◯◯ 材料

黑色部分
馬蹄粉 80 克
粘米粉 20 克
糖 100 克
黑芝麻醬 3 湯匙
水 2.5 杯

◯◯ 白色部分
馬蹄粉 80 克
粘米粉 20 克
杏仁霜 4 湯匙
糖 120 克
鮮奶 3/4 杯
水 2.5 杯

◎ 做法

黑色部分

1. 將馬蹄粉、粘米粉、黑芝麻醬和 3/4 杯水調勻，用篩過濾成黑芝麻水。

2. 糖與 2.5 杯水放入鍋內，煮滾，加入 1/4 黑芝麻水，拌成稀糊待涼，再加入餘下的黑芝麻水拌勻成黑色部分。

白色部分

1. 馬蹄粉、粘米粉、鮮奶和杏仁霜拌勻，用篩過濾成杏仁奶。

2. 糖與 2.5 杯水放入鍋內，煮滾，加入 1/4 杏仁奶拌成糊，待涼，再加入餘下的杏仁奶拌勻成白色部分。

◎ 綜合做法

蒸盤抹上油，倒入適量黑色粉漿蒸 5 分鐘，再倒入等量的白色粉漿再蒸 5 分鐘，如此類推，最後一層蒸 15 分鐘即可。

◎ 零失敗技巧 ◎
Successful cooking skills

如何做出外形美觀的黑白芝麻千層糕？

蒸每層糕時，必須倒入等量之粉漿，做成每層相等之厚度。

How do you make the layered cake look perfect?

Before steaming each layer, you must pour on the same amount so that each layer will be of the same thickness.

軟滑之千層糕有何竅門？

無論是黑色或白色粉漿，調勻後必須用隔篩過濾，才能做出幼滑之千層糕。

Is there any secret trick to make the layered cake silky smooth?

Make sure you pass the batter, no matter white or black, through a fine wire mesh before steaming. That would make the cake velvety.

Ingredients

Black sesame layer
80 g water chestnut starch
20 g long-grain rice flour
100 g sugar
3 tbsp black sesame spread
2.5 cups water

White almond milk layer
80 g water chestnut starch
20 g long-grain rice flour
4 tbsp almond milk powder
120 g sugar
3/4 cup milk
2.5 cups water

Method

Black sesame layer

1. Mix together water chestnut starch, rice flour, black sesame spread and 3/4 cup of water. Stir well. Strain and save the black sesame extract.

2. In a pot, put in 2.5 cups of water and sugar. Bring to the boil. Stir in 1/4 of the strained black sesame extract from step 1. Stir well into a thin paste. Leave it to cool. Pour in the remaining black sesame extract. Stir again.

White almond milk layer

1. Mix together water chestnut starch, rice flour, milk and almond milk powder. Stir well. Strain and save the almond milk mixture.

2. In a pot, put in 2.5 cups of water and sugar. Bring to the boil. Stir in 1/4 of the almond milk mixture. Leave it to cool. Pour in the remaining almond milk mixture. Stir well.

Assembly

Grease a steaming tray. Pour in a thin layer of black sesame batter. Steam for 5 minutes. Pour in about the same amount of almond milk batter. Steam for 5 minutes. Pour in the black and the white batter in alternate manner until the tray is full. Steam for 15 minutes at last.

黑芝麻杏仁千層糕

鳳梨酥卷

Crumbly Pineapple Pastry Rolls

容易指數：⭐⭐☆☆☆

◎ 鳳梨餡料
新鮮菠蘿 1 個或菠蘿 1 罐
糖 2 湯匙
麥芽糖 2 湯匙
植物牛油 2 茶匙
肉桂 1 枝
丁香 3 至 4 粒

◎ 皮料
牛油 60 克
麵粉 80 克
吉士粉 1 湯匙
奶粉 1 湯匙
糖霜 30 克

◎ 做法

餡料

1. 菠蘿切碎,擠去多餘水分。
2. 菠蘿碎、糖、麥芽糖、丁香、肉桂放入鍋內,用慢火煮至餡料濃稠可推起,盛起,棄去丁香、肉桂;加入植物牛油搓勻,可搓成湯圓狀。

皮料

1. 麵粉、吉士粉及奶粉一起篩勻。
2. 牛油與糖霜打透。
3. 加入麵粉拌勻成粉糰,放入雪櫃雪至略硬。

◎ 綜合做法

1. 將粉糰分成數粒,然後將一粒粉糰放入專做鳳梨酥卷的模型內,擠出狗牙狀長條,放上適量餡料捲好,收口向下,冷藏片刻。
2. 放入已預熱 190℃焗爐,焗約 20 分鐘即成。

◎ Ingredients

Pineapple filling

1 can or 1 whole fresh pineapple
2 tbsp sugar
2 tbsp maltose syrup
2 tsp margarine
1 cinnamon stick
3-4 cloves

◎ Crust

60 g butter
80 g flour
1 tbsp custard powder
1 tbsp milk powder
30 g icing sugar

◎ Method

Filling

1. Finely dice the pineapple. Squeeze out excess moisture.
2. Put pineapple, sugar, maltose syrup, cloves and cinnamon stick into a pot. Cook over low heat until thick and jam-like consistency. Remove from heat and set aside. Discard cinnamon sticks and cloves. Add margarine and mix well. Roll into small balls.

1

2

3

Crust

1. Sieve flour, custard powder and milk powder together.

2. Beat butter with icing sugar until well incorporated.

3. Add flour and mix well. Refrigerate the dough until firm.

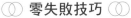 Assembly

1. Divide the dough into a few pieces. Put one piece into a dough press with the special die. Press to make an indented long strip of dough. Put some filling on it and roll the dough around the filling. Place it with the seam facing down. Refrigerate briefly.

2. Bake in a preheated oven at 190°C for about 20 minutes. Serve.

零失敗技巧
Successful cooking skills

用罐裝或新鮮菠蘿炮製，哪款味道最佳？
用新鮮菠蘿炮製「鳳梨酥卷」，味道更加鮮美，菠蘿味濃重。
May I use canned pineapple or fresh pineapple to do it?
For the best result, use fresh pineapple as the filling.

哪裏購買專做鳳梨酥卷的模型？
上海街專門售賣入廚用具的店舖，或烘焙食品店均有售。
Where can I find the mould for making pineapple rolls?
You can get them in kitchen utensil stores on Shanghai Street in Hong Kong. You may also try baking supply stores.

為何皮料拌勻後需要冷藏？
因麵糰含有牛油，冷藏一會以免牛油經混合後溶解，令麵糰溶掉。
Why do you refrigerate the pastry dough after mixing?
The pastry dough contains butter. I refrigerate it so that the butter stays solid without melting. If the butter melts, the pastry dough will be too soft and difficult to work with.

新手入廚
做好餸
Skilful recipes for novice cooks

編者 Forms Kitchen編輯委員會	Editor Editorial Committee, Forms Kitchen
美術設計 馮景蕊	Design Carol Fung
排版 辛紅梅	Typography Cindy Xin
出版者 香港鰂魚涌英皇道1065號 東達中心1305室 電話 傳真 電郵 網址	Publisher Forms Kitchen Room 1305, Eastern Centre, 1065 King's Road, Quarry Bay, Hong Kong. Tel: 2564 7511 Fax: 2565 5539 Email: info@wanlibk.com Web Site: http://www.wanlibk.com 　　　　　http://www.facebook.com/wanlibk
發行者 香港聯合書刊物流有限公司 香港新界大埔汀麗路36號 中華商務印刷大廈3字樓 電話 傳真 電郵	Distributor SUP Publishing Logistics (HK) Ltd. 3/F., C&C Building, 36 Ting Lai Road, Tai Po, N.T., Hong Kong Tel: 2150 2100 Fax: 2407 3062 Email: info@suplogistics.com.hk
承印者 中華商務彩色印刷有限公司	Printer C & C Offset Printing Co., Ltd.
出版日期 二零一九年三月第一次印刷	Publishing Date First print in March 2019

鳴謝以下作者提供食譜（排名不分先後）：
黃美鳳、Feliz Chan、Winnie姐